Elementary Atomic Structure

ELEMENTARY ATOMIC STRUCTURE

SECOND EDITION

G. K. WOODGATE

CLARENDON PRESS · OXFORD

Oxford University Press, Walton Street, Oxford OX2 6DP

Oxford New York Toronto
Delhi Bombay Calcutta Madras Karachi
Petaling Jaya Singapore Hong Kong Tokyo
Nairobi Dar es Salaam Cape Town
Melbourne Auckland

and associated companies in
Beirut Berlin Ibadan Nicosia

Oxford is a trade mark of Oxford University Press

Published in the United States
by Oxford University Press, New York

British Library Cataloguing in Publication Data
Woodgate Gordon Kemble
Elementary atomic structure.—2nd ed.
1. Atoms
I. Title
539.14   QC173   79–41673
ISBN 0–19–851156–6

Printed in Hong Kong

Preface to the first edition

I have given this book the title *Elementary Atomic Structure* to indicate that it is neither introductory nor advanced. It goes into the subject of atomic structure at an elementary level in somewhat more detail than is to be found in most of the large books on atomic physics.

The book is an extended version of a set of lectures given to undergraduates at Oxford. By the time they have reached their third year these undergraduates have had courses in atomic physics and in quantum mechanics. While I regard an introductory course in atomic physics as a prerequisite, I have tried to write the book in such a way that the earlier parts can be given a first reading concurrently with a study of formal quantum mechanics.

The character of the book has been determined by two rather different considerations: a desire to instil a feeling for orders of magnitude, which I regret to say is too often lacking, and a desire for brevity. I have tried to use real examples, with numbers, to illustrate the application of quantum mechanics to atomic structure, and although these examples are elementary, I hope they show that real life is hedged about with approximations which one should attempt to understand. As for the second consideration—the book is meant to be short enough for the undergraduate to feel that he will actually be able to read it, do the problems, *and* read other books in the time that he can afford to devote to a study of atomic structure.

Of course, the shortness of the book has led to a selection of material. In particular, I have omitted the usual discussion of the width of lines in optical spectroscopy. I draw attention to this here, for I feel that nowadays a treatment of this topic should be combined with a discussion of measurements in radiofrequency spectroscopy and in modern methods of studying lifetimes and line shapes. Pressure broadening alone, for example, is a large subject of an advanced nature. In general I have avoided discussing the interaction of atoms with their environment, except for the interaction with applied electromagnetic fields. I have also regarded the Dirac equation as being beyond the scope of this book because at the present time a study of this topic marks the beginning of graduate work in most universities. It may be tantalizing for the better undergraduate that, as a result, I have had to *quote* the relativistic contributions to the fine structure of hydrogen in chapter 4, but perhaps he may be stimulated thereby to explore further. What I have attempted is an elementary discourse, proceeding as logically as possible, on the hierarchy of interactions which are of importance in describing the internal structure of atoms. It is natural to treat these interactions at each stage by perturbation theory.

The problems are intended to be an important part of the book, complementary to the main text. For example, in chapter 5 variational techniques get a brief mention in the text, but in problem 5.2 the reader is asked to do the simplest variational calculation for himself and is given guidance on how to proceed. There is, of course, no substitute for doing problems.

Apart from emphasizing at an early stage in the book that a prior or concurrent study of quantum mechanics is necessary, I have given a few references in footnotes to encourage wider reading of an immediate nature, but I have not thought it appropriate to aim at a large bibliography in a book of this size. The undergraduate will find adequate references in *Atomic Spectra* by H. Kuhn (Second Edition, Longmans, 1970), a book which I recommend also for its wealth of spectroscopic information.

I hope that the book is sufficiently open-ended in its more advanced passages that the undergraduate, in progressing beyond the elementary level, will find himself prepared both to meet the challenge of the advanced texts such as *The Theory of Atomic Spectra* by E. U. Condon and G. H. Shortley (Cambridge University Press, 1951) and *Quantum Theory of Atomic Structure* by J. C. Slater (McGraw-Hill, 1960), and to welcome the use of modern techniques in the calculation of atomic structures when he first comes across them.

Finally, I should like to thank those colleagues and pupils who have kindly agreed to take a look at parts of the manuscript. I should also like to express my gratitude to Dr Kuhn who first introduced me to the subject of atomic structure and who encouraged me in the study of it.

G. K. WOODGATE

Preface to the second edition

In this SI edition I have kept the general content of the book the same: it is neither much longer nor more advanced than the first edition. I have taken the opportunity, however, to correct all the errors I could find and to change those passages of the book which I had come to feel were either downright wrong or badly expressed. I have also added some more problems, particularly of the step-by-step kind which supplement the text, and which experience has shown to be the most fruitful for teaching purposes.

Several people, amongst whom I mention Professor P. G. H. Sandars, Mr C. W. P. Palmer and Dr K. R. Lea, have given me helpful advice in the preparation of this new edition and I am grateful to them. I should like to thank especially Dr G. A. Brooker for taking the trouble to make a thorough criticism, based on his teaching experience, of parts of the book.

Contents

Contents

1. Introduction

The structure of atoms is described successfully by the theory of quantum mechanics, and there is no doubt that in order to understand atomic structure one has to learn quantum mechanics. However, quantum mechanics came into being as the culmination of the development of earlier theories. Of these, Bohr's theory of hydrogen, based on Rutherford's nuclear atom and incorporating the ideas of Planck, was the famous starting point for atomic structure. Bohr's semi-classical theory was not general enough to describe more than the gross features of the simplest one-electron atom, but it provided a *model* of an atom which is easily visualized. Because of its vividness we tend to retain Bohr's picture in the back of the mind, although we reject the classical mechanistic parts of Bohr's model as a serious explanation of atomic structure.

In this book we shall use elementary quantum mechanics to describe the structure of atoms. But we shall also come across other models which, for reasons to be explained later, have a greater claim to validity than Bohr's picture of a hydrogen atom.

Bohr contributed much more than a mere semi-classical model. For the interpretation of atomic spectra the main point of his theory was the concept of stationary states of given energy, together with the frequency condition

$$h\nu_{21} = E_2 - E_1 \qquad (1.1)$$

which expresses the conservation of energy in a radiative transition of frequency ν_{21} between stationary states of energies E_1 and E_2 (h is Planck's constant). We take these ideas very seriously but, in accepting the postulates of quantum mechanics as having a validity much wider than their application to atomic structure, we find that Bohr's powerful ideas are included as special results of a more general theory. As always, such an attitude is aesthetically more satisfactory and scientifically more fruitful.

1.1. Empirical aspects

Theories of atomic structure grew up as attempts to explain the *spectra* of free atoms, that is the electromagnetic radiation emitted or absorbed by

1

atoms. Empirical analysis of the discrete frequencies found in the spectrum of an atom was facilitated by the *Rydberg–Ritz combination principle* which stated in effect that the frequencies v (or rather the wavenumbers $\tilde{v} = v/c$ where c is the speed of light) could be expressed as the differences of *terms*:

$$\tilde{v}_{ji} = T_i - T_j. \tag{1.2}$$

The consequence of this is that if $\tilde{v}_{ji}$ and $\tilde{v}_{ki}$ occur in the spectrum, so might $\tilde{v}_{kj}$ where $\tilde{v}_{kj} = \tilde{v}_{ki} - \tilde{v}_{ji}$. By listing an array of frequencies and their differences one can assign, for a simple spectrum, a set of term values consistent with all the observed frequencies. For example, one would start by fixing the terms T_k and T_j on the basis that the difference $\tilde{v}_{ki} - \tilde{v}_{ji}$ occurs for several i in the array. In complex spectra, consisting of hundreds of thousands of lines, such frequency differences often occur fortuitously: hence the need for high precision in the measurements. In fact many complex spectra, in particular those of the rare earths, have not been analysed.

The term values T_i are written as positive numbers in units of cm^{-1}, recently re-named the Kayser (K). The new name is more commonly found when the sub-unit milli-Kayser is used: $1\ mK = 10^{-3}\ cm^{-1}$. The energies E_i in eq. (1.1) are negative numbers, and there is an obvious relationship between eq. (1.1) and (1.2) provided

$$T_i = -E_i/hc. \tag{1.3}$$

Thus term values are equivalent to binding energies whose zero is the ionization limit of the atom. Modern tables of energy levels† adopt a different ordering: the most negative energy level, the ground state, is taken as the zero of energy and the excited states have positive energies above the ground state up to a maximum at the ionization limit.

In the simple spectra there are striking regularities. The frequencies appear in definite series and so, by the combination principle, the energy levels occur in series which can sometimes be expressed in simple analytical form. For example, for hydrogen-like atoms

$$E_n = -hcR/n^2, \tag{1.4}$$

and for alkali-like atoms

$$E_n = -hcR/(n - \delta)^2. \tag{1.5}$$

In these formulae R is a constant for a given atom, the index number n is a running integer for a given series, and δ is a constant, not necessarily

---

† For example: *Atomic Energy Levels*, Vols. I, II, and III by Charlotte E. Moore. National Bureau of Standards Circular Number 467. It cannot be too strongly emphasized that a feel for numbers, as derived from a study of such tables, is all-important as a basis for understanding what the theory is about.

integral, for a given series. These empirical formulae fit the spectra under conditions of low resolution. Equation (1.5) is less accurate than (1.4). The index number n is the *principal quantum number* of Bohr's theory and of quantum mechanics, although in some old energy-level diagrams which the reader may come across the ground state is labelled always by $n = 1$ which does not correspond to the labelling by principal quantum number.† Another label for distinguishing different series in empirical spectroscopy was the integer L (or l in single-electron spectra) which corresponds to the orbital angular momentum quantum number. It is denoted by the letter-code

$$l = 0 \quad 1 \quad 2 \quad 3 \quad 4 \quad 5 \ldots$$
$$\quad S \quad P \quad D \quad F \quad G \quad H \ldots$$

In describing a transition between two energy levels, each labelled by nl, the convention is to write the *lower* level first: thus the (unresolved) yellow line of sodium is described by 3S—3P. This is the lowest-frequency member of the so-called *principal series* 3S—nP.

Apart from frequency, another quantity which can be measured for spectral lines is their relative intensity. A line which might be expected to fit into an array of frequencies on the basis of the combination principle may actually be absent. A zero intensity (a forbidden transition) is interpreted theoretically in terms of a *selection rule*.

It is the purpose of the theory to describe in a consistent manner the features of atomic structure by which the energy levels and intensities of transitions are explained quantitatively. In the course of the development of quantum mechanics for this purpose, many problems in the much wider field of pure physics and natural philosophy have been elucidated. There is also a sense in which the theory of atomic structure is an applied science: quantitative calculations of the behaviour of free atoms are required as a basis for the less well-defined fields (that is, fields in which the experiments are less easily controlled) of, for example, solid-state physics, plasma physics, and—an extreme case—astrophysics.

1.2. Orders of magnitude

We have outlined so far the empirical aspects of what is called the gross structure of the energy levels of an atom. This is to be described theoretically in terms of the largest interaction within an atom, that is by a potential energy representing the largest force. Under higher and higher resolution it is found that the energy levels split into sub-levels with progressively smaller separations. These splittings can be interpreted in terms of smaller and smaller interactions by approximation methods. The chief

† For example, the principal quantum number describing the ground state of sodium is $n = 3$ and for caesium it is $n = 6$, but the index number used in the old type of diagram may nevertheless be $n = 1$ in both cases.

method employed is that of perturbation theory. It is really very fortunate that this can be done at all.

It is of vital importance to have a clear quantitative picture of this hierarchy of splittings in order to have a feel for each interaction in its proper perspective. We shall now take an overall view of the orders of magnitude of the more important interactions. We shall find that we are dealing entirely with electromagnetic interactions.

(a) The electrostatic interaction between a heavy nucleus of charge Ze and an electron of charge $-e$, mass m_0, is of the form† $-Ze^2/4\pi\varepsilon_0 r$ where r is the distance between nucleus and electron. This term leads to an energy of the ground state of a hydrogen-like atom, E_1, of order

$$E_1 = -\frac{1}{(4\pi\varepsilon_0)^2}\frac{m_0 Z^2 e^4}{2\hbar^2} \approx -110,000 \text{ cm}^{-1} \approx -14 \text{ eV}, \quad (1.6)$$

where $\hbar$ is Planck's constant divided by 2π. Alternatively, we can express E_1 in terms of a length:

$$E_1 = -\frac{1}{4\pi\varepsilon_0}\frac{Z^2 e^2}{2a_0} = -\frac{1}{4\pi\varepsilon_0}\frac{Ze^2}{2(a_0/Z)} \quad (1.7)$$

where $a_0 = 4\pi\varepsilon_0\hbar^2/(m_0 e^2) \approx 0.5 \times 10^{-10}$m. The quantity a_0 is called the radius of the first Bohr orbit and is a measure of the size of atoms. It is also a measure of the domain in which quantum effects are dominant: for if an electron in a one-electron atom is regarded as being contained in a box of size a_0/Z, with kinetic energy $\overline{p^2}/2m_0 = Z^2 e^2/8\pi\varepsilon_0 a_0$ according to Bohr's semi-classical picture, then its mean momentum $\bar{p}$ is zero (otherwise it would not be contained in the box) and the uncertainty in its momentum, defined by

$$\Delta p \equiv (\overline{p^2} - \bar{p}^2)^{1/2}, \quad (1.8)$$

becomes

$$\Delta p = (\overline{p^2})^{1/2} = \frac{\hbar}{(a_0/Z)}. \quad (1.9)$$

By the Uncertainty Principle, the minimum uncertainty in the position of the electron is then a_0/Z, the size of the box. This certainly represents quantum behaviour.

Another important quantity corresponds semi-classically to the speed of the electron in the first Bohr orbit as a fraction of the speed of light:

$$p_1/(m_0 c) = \frac{\hbar Z}{m_0 c a_0} = \frac{1}{4\pi\varepsilon_0}\frac{Ze^2}{\hbar c} = Z\alpha \quad (1.10)$$

where $\alpha = e^2/(4\pi\varepsilon_0\hbar c) \approx 1/137$ is called the fine-structure constant. This dimensionless constant is a measure of the strength of electromagnetic

† SI units are used throughout the book. See appendix E.

forces. Because of its magnitude, factors like $(1 - Z^2\alpha^2)^{1/2}$ which will occur as relativistic corrections can obviously be expanded binomially and will give good approximations in zeroth and first order. Also one observes that the electrostatic binding energy $(-E_1)$ of the electron in a hydrogen-like atom is of the order of a factor $Z^2\alpha^2$ less than the rest energy of the electron:

$$-E_1 = \tfrac{1}{2}Z^2\alpha^2 m_0 c^2. \tag{1.11}$$

(b) In many-electron atoms the electrostatic interaction with the nucleus is summed over all the electrons to contribute a term

$$\sum_i (-Ze^2/4\pi\varepsilon_0 r_i)$$

in the potential energy. In addition there are electrostatic repulsion terms between each pair of electrons:

$$\sum_{i>j} (e^2/4\pi\varepsilon_0 r_{ij}).$$

While the latter is in general too large to be treated as a small perturbation on the former, part of the mutual repulsion between electrons is directed away from the nucleus and cancels part of the nuclear attraction. One can then think of valence electrons moving in a net *central field* with their motion perturbed by the non-central part of the electrostatic forces. The central field is the dominant effect and leads to a classification of the gross structure of the energy levels of an atom by *configurations*. Different configurations of a neutral atom have energy separations up to about 30,000 cm^{-1}, or about 4 eV. The residual non-central electrostatic interaction between electrons is a smaller effect and causes each configuration to consist of a number of *terms* whose energy separations may typically be of the order of 3,000 cm^{-1}. That is to say, in the central-field approximation each configuration really consists of several terms all of the same energy, but in the higher approximation of including the non-central interaction each term of a configuration has a different energy. When the approximation of treating the residual electrostatic interaction as a *small* perturbation is inadequate, one speaks of *configuration mixing*. For *simple* spectra the central-field approximation is a very good one because the disposition in spherically symmetrical *shells* of all but the one or two valence electrons leads to a great simplification. For *complex* spectra the approximation begins to break down. This discussion of the electrostatic interactions is the least tidy aspect of our attempt to describe an idealized hierarchy of splitting of energy levels.

(c) The next interaction in order of magnitude is a relativistic one involving *electron spin*. The *spin–orbit* interaction is the largest relativistic effect and is responsible for *fine structure*. Each *term* splits into *levels* whose separations are of the order of 1–$1,000$ cm^{-1}, depending strongly on the atomic number Z. The interaction energy is of the form

$$V = -\boldsymbol{\mu} \cdot \mathbf{B}_{el} \tag{1.12}$$

5

where $\mathbf{B}_{el}$ is the internal magnetic field at an electron whose intrinsic magnetic momentum is $\boldsymbol{\mu}$. The size of $\boldsymbol{\mu}$ is of the order of a Bohr magneton:

$$\mu_B = e\hbar/(2m_0) \approx 9\cdot3 \times 10^{-24}\,\mathrm{JT}^{-1}. \tag{1.13}$$

B_{el} for a hydrogen-like atom is of the order of $(\mu_0/4\pi)Z\mu_B/r^3 \approx (\mu_0/4\pi)$ $Z\mu_B/(a_0/Z)^3$, so the energy shift ΔE is

$$\Delta E \approx \frac{-\mu_0}{4\pi}\frac{\mu_B^2 Z^4}{a_0^3} = \tfrac{1}{2}Z^2\alpha^2\left(-\frac{1}{4\pi\varepsilon_0}\frac{Z^2 e^2}{2a_0}\right) = \tfrac{1}{2}Z^2\alpha^2\,E_{gross}. \tag{1.14}$$

Thus the fine structure is smaller than the gross structure by a factor $Z^2\alpha^2$. Expression (1.14) shows why α is called the fine-structure constant. From (1.13) and the definition of a_0 we have also

$$\mu_B = \tfrac{1}{2}\alpha c(ea_0), \tag{1.15}$$

showing that atomic magnetic dipole moments are smaller than atomic electric dipole moments by a factor $\alpha c/2$. The range of fine structures, $1-1{,}000\ \mathrm{cm}^{-1}$, quoted above corresponds to internal magnetic fields in the range $10-10^4$ T.

(d) The *levels* are split further into *states* by the application of a laboratory magnetic field, B_{lab}. In practice $B_{lab} \leqslant 1\mathrm{T}$, so the splitting is of the order of

$$\mu_B\,B_{lab} \leqslant 1\ \mathrm{cm}^{-1}.$$

This is called the Zeeman effect.

(e) Whereas the nuclear electric charge, Ze, is responsible for the largest interaction in which the electrons take part, the other nuclear multipole moments—magnetic dipole, electric quadrupole, etc.—give rise to the smallest interaction so far to be considered. This is called hyperfine structure. For example, nuclear magnetic dipole moments are of the order of a nuclear magneton $\mu_N = e\hbar/(2M)$ where M is the proton mass. Since $M = 1{,}836\,m_0$, $\mu_N = \mu_B/1{,}836$ and the interaction of the nuclear magnetic moment with the magnetic field produced at the nucleus by the electrons is of the order of

$$-\mu_N\,B_{el} = -(1/1{,}836)\,\mu_B\,B_{el} \sim 0\cdot001-1\ \mathrm{cm}^{-1}. \tag{1.16}$$

Thus the magnetic hyperfine structure is less than the fine structure by a factor of about m_0/M. One could write the interaction (1.16) as $-\mu_B B_N$ where B_N is the magnetic field produced at the electrons by the nuclear magnetic moment. Then $B_N \approx (m_0/M)B_{el} \sim 10^{-2}-10\mathrm{T}$. Whereas the spectral lines corresponding to the gross energy differences lie in the visible and ultraviolet and are examined optically, hyperfine structure splittings are in the radiofrequency region and radiofrequency methods have naturally come to be used in their investigation, especially since the technical means of doing so have become available.

(f) We can see at once that the Zeeman effect of hyperfine structure, in which a laboratory field B_{lab} ($\lesssim 1T$) is applied, gives a splitting of hyperfine structure levels which is not necessarily small compared with the hyperfine structure splitting itself: $B_{lab} \nless B_N$.

In table 1.1 below we summarize the orders of magnitude of all these interactions in several systems of units, including an effective temperature unit defined by $TK = h\nu/k$, where k is Boltzmann's constant.

Table 1.1. Summary of orders of magnitude

Interaction	Magnitude			
	cm^{-1}	eV	Hz	K
(a) Central electrostatic	30,000	4	10^{15}	43,000
(b) Residual electrostatic	3,000	0·4	10^{14}	4,300
(c) Fine structure	1–1,000	10^{-4}–10^{-1}	3×10^{10}–3×10^{13}	1·4–1,400
(d) Zeeman effect of fine structure	1	10^{-4}	3×10^{10}	1·4
(e) Hyperfine structure	10^{-3}–1	10^{-7}–10^{-4}	3×10^7–3×10^{10}	$1·4 \times 10^{-3}$–1·4

Of course the scheme outline in table 1.1 is not always as clear-cut as we have made out. Sometimes there is much more overlapping of the categories than has been indicated. But as a starting point for the application of the perturbation method the scheme is often realistic enough. In the rest of this book we shall be concerned with the interactions, one by one, mentioned in this survey.

Problems

1.1 In the table below are given eight of the wavenumbers $\tilde{\nu}_n$ for the transitions 3S—nP in sodium. Devise an extrapolation procedure to find the ionization limit of sodium with a precision justified by the data. Convert the result into Ångstrom units and into electron volts.

n	$\tilde{\nu}\ cm^{-1}$	n	$\tilde{\nu}\ cm^{-1}$
7	38,541	11	40,383
8	39,299	12	40,566
9	39,795	13	40,706
10	40,137	14	40,814

1.2. Evaluate the ratio of the gravitational to the electrostatic force of attraction between a proton and an electron.

1.3 For thermal equilibrium at temperature T an appropriate measure of energy is kT where k is Boltzmann's constant. Convert the following into units of K: 1 Rydberg; $10^3\ cm^{-1}$, $1\ cm^{-1}$, $10^{-3}cm^{-1}$; 1 eV.

1.4. The 3P term of sodium lies approximately 17,000 cm^{-1} above the ground term 3S. The relative statistical weights of the 3P and 3S terms are

$g(3P)/g(3S) = 6/2 = 3$. If a bulb of sodium vapour is in thermal equilibrium with a temperature bath at TK, at what temperature would 1 per cent of the atoms be in the 3P term? Assume that all other terms lie much higher than the 3P term.

1.5 The ground level of an atom is split into two Zeeman states of equal statistical weight separated by 10,000 MHz. An assembly of such atoms is in thermal equilibrium at temperature TK. What is the fractional population difference of the Zeeman states when $T = $ 300K, 20K, 4K, 1·5K?

2. The hydrogen atom: gross structure

Hydrogen, with only one electron, is the simplest possible atomic system. Moreover, the central electrostatic interaction between electron and nucleus, which is the only interaction considered in this chapter, is represented by the potential energy $V(r) = -Ze^2/4\pi\varepsilon_0 r$. This Coulomb interaction is of a special form with respect to its radial dependence ($V(r) \propto r^{-k}$ where $k = 1$) and leads to special results which are by no means typical of atoms with one valence electron and a core of spherically symmetrical electron shells. The disadvantage of treating hydrogen first (as is done here and in most other textbooks) is that the student might be misled into thinking that the detailed results are typical of other atoms. This disadvantage is partly compensated by the fact that the problem can be solved exactly within the framework of the initial physical assumptions, and the solution exemplifies the methods of quantum mechanics. For this same reason, of course, hydrogen is one of the best testing grounds for the theory and much fundamental work has been done on it. The nature of, and necessity for, approximation methods becomes apparent later.

In the next three sections we give, in outline only, a discussion of Schrödinger's equation and of the postulates of wave mechanics in order to summarize the concepts which we shall need to use. The remarks in these sections are intended to serve as a reminder to the reader about what he should know of the elementary theory. If he feels that his background is inadequate at this stage he is urged to study concurrently a book† on quantum mechanics, for we shall come to the point without much preamble.

2.1. The Schrödinger equation

Let us take the classical equation of conservation of energy

$$E = \frac{\mathbf{p}^2}{2m_0} + V(x, y, z) \tag{2.1}$$

† An attempt has been made to keep the notation consistent with that of R. H. Dicke and J. P. Wittke, *Introduction to Quantum Mechanics*, Addison-Wesley, 1960.

for the motion of a particle of mass m_0 and momentum $\mathbf{p}$ moving in a potential V which is not an explicit function of time t, and make the replacements

$$E \rightarrow i\hbar \frac{\partial}{\partial t},$$

$$\mathbf{p} \rightarrow -i\hbar \, \nabla. \tag{2.2}$$

If these differential operators are allowed to operate on a wave function $\Psi = \Psi(x, y, z, t)$ we obtain a differential equation

$$i\hbar \frac{\partial \Psi}{\partial t} = -\frac{\hbar^2}{2m_0} \nabla^2 \Psi + V\Psi, \tag{2.3}$$

or

$$i\hbar \frac{\partial \Psi}{\partial t} = \mathscr{H}\Psi \tag{2.4}$$

where

$$\mathscr{H} = \frac{-\hbar^2 \nabla^2}{2m_0} + V$$

is called the Hamiltonian (for a conservative system). Here $\mathscr{H}$ is an operator representing the sum of the kinetic and potential energies.

The operator relacements (2.2), together with the existence of a wave function Ψ, may be regarded as the fundamental postulates of wave mechanics.

The wave equation (2.3) is called Schrödinger's equation for the motion of a particle in a potential $V(x, y, z)$. This equation is linear in Ψ which is a *complex* wave amplitude. The generally accepted interpretation of Ψ is that proposed by Born: if Ψ^* is the complex conjugate of Ψ, $\Psi^*\Psi$ represents a probability density for finding the particle at the co-ordinates (x, y, z, t) as the result of a measurement of its position at time t. The linearity of Schrödinger's equation implies that if Ψ_1 and Ψ_2 are solutions of the equation then so is $c_1\Psi_1 + c_2\Psi_2$, where c_1 and c_2 are constants. This *superposition principle* is of vital importance because it gives rise to wave-mechanical interference which is the fundamental idea behind our understanding of quantum-mechanical phenomena. Indeed, other wave equations, notably Dirac's relativistic wave equation, are constructed in such a way as to retain this feature of linearity.

There can be no pretence that the introduction of Schrödinger's equation given above is anything but mysterious. But then the initial postulates of a theory, by their nature, are not derived from anything else. One can only present them in the most plausible form and find justification for them in the validity of their application to physical problems. The classical

2.1. The Schrödinger equation

form of the Hamiltonian, eq. (2.1), is still apparent in Schrödinger's equation, despite the disguise in operator form, eq. (2.2). We shall have occasion to specify particular problems by expressing the potential energy largely in classical terms. Thus we may still recognize in the solution of the problem certain classical features and we may be able to build a model, or classical mental picture, based on these features. But the other aspect of Schrödinger's equation, namely the postulation of a wave function, will have to be understood in a more abstract manner. In particular we shall have difficulty in visualizing the interference phenomena mentioned above except in a mathematical sense.

Ψ, the amplitude describing the motion of a quantum-mechanical 'particle', is analogous to the amplitude, ϕ, describing the motion of a wave, provided that we make use of the ideas of Einstein and de Broglie in transferring from the wave concept to the particle concept. For a plane wave, for example, of definite angular frequency ω and wave vector $\mathbf{k}$ ($|\mathbf{k}| = 2\pi/\lambda$ where λ is the wavelength)

$$\phi = \phi_0 \, e^{i(\mathbf{k} \cdot \mathbf{r} - \omega t)}. \tag{2.5}$$

With the replacements

$$E = \hbar\omega \quad \text{(Einstein)} \tag{2.6}$$

and

$$\mathbf{p} = \hbar\mathbf{k} \quad \text{(de Broglie)} \tag{2.7}$$

we arrive at the amplitude Ψ for a free particle of definite energy E and momentum $\mathbf{p}$

$$\Psi = \Psi_0 \, e^{i/\hbar(\mathbf{p} \cdot \mathbf{r} - Et)}. \tag{2.8}$$

The operations $i\hbar \, (\partial/\partial t)$ and $-i\hbar\nabla$ give

$$i\hbar \frac{\partial}{\partial t} \Psi = E\Psi \tag{2.9}$$

and

$$-i\hbar \, \nabla\Psi = \mathbf{p}\Psi, \tag{2.10}$$

which are *eigenvalue* equations. Thus the postulates (2.2) about operators representing the classical variables E and $\mathbf{p}$ of a particle are consistent with the quantum postulates of Einstein and de Broglie. The wave equation of which ϕ is a solution is of second order in t provided $\omega \propto |\mathbf{k}|$, which is the condition for classical waves to be non-dispersive. On the other hand the 'matter wave' describing the motion of a free particle has $E \propto \mathbf{p}^2$, or, from eqs. (2.6) and (2.7), $\omega \propto |\mathbf{k}|^2$. This essential distinction between a matter wave and a classical non-dispersive wave is embodied in the matter wave equation: Schrödinger's equation is only of first order in t.

In Schrödinger's equation Ψ is a function in co-ordinate space (x, y, z) and time. This is only a particular way of working. One could form a function Φ in momentum space (p_x, p_y, p_z) and time. Ψ and Φ are Fourier transforms of each other, and in the momentum representation the operator replacements for classical variables are different from those (2.2) which we have postulated in the co-ordinate representation. In fact Leighton† in his book (p.93ff.) adopts the Fourier relationship between Ψ and Φ as a postulate, and finds the operator replacements (2.2) as a special result applicable in co-ordinate space. The reason why one nearly always uses the function Ψ in describing atomic structure is because the potential V is easily expressed as a function of x, y, and z.

Schrödinger's equation is equivalent to the non-relativistic form of the classical equation of conservation of energy (2.1). It therefore leads to a *non-relativistic* wave-mechanical theory. However, if we consider the four-vector displacement (x, y, z, ict) and the four-vector momentum $(p_x, p_y, p_z, iE/c)$ we see that the recipe for converting a momentum component to its operator form gives, for the fourth component,

$$iE/c \rightarrow -i\hbar \frac{\partial}{\partial(ict)} = \frac{i}{c}\left(i\hbar \frac{\partial}{\partial t}\right)$$

or

$$E \rightarrow i\hbar \frac{\partial}{\partial t}, \tag{2.11}$$

which is the same as the postulate made separately about E (as distinct from **p**) in (2.2). The construction of a relativistic wave equation for a free particle from the relativistic equation of conservation of energy

$$E^2 = p^2c^2 + m_0^2c^4 \tag{2.12}$$

follows from the recipe for operators. Thus:

$$\frac{1}{c^2}\frac{\partial^2 \Psi}{\partial t^2} = \nabla^2 \Psi - \frac{m_0^2 c^2}{\hbar^2}\Psi. \tag{2.13}$$

This is called the Klein–Gordon equation for a free particle without spin. Electron spin is not a classical concept, nor is it a wave-mechanical concept in x, y, z, t space, and its introduction into a wave equation is a much more complicated matter. This was achieved by Dirac. Notice that eq. (2.13) naturally contains the speed of light, c, whereas the non-relativistic Schrödinger equation does not.

The point about this discussion of four-vectors is that functions of x and of p_x (or rather k_x) are related by a Fourier transformation in classical physics. The same applies to t and E (or rather ω). These pairs of variables

† R. B. Leighton, *Principles of Modern Physics*, McGraw-Hill, 1959.

are conjugate variables in classical mechanics and are just the pairs which, in their operator form, do not commute:

$$xp_x - p_xx \rightarrow x\left(-i\hbar\frac{\partial}{\partial x}\right) -\left(-i\hbar\frac{\partial}{\partial x}\right)x = i\hbar\cdot \tag{2.14}$$

These pairs are also connected by Heisenberg's Uncertainty Principle which is another realization of the properties of non-commuting variables. The Fourier relationship between Ψ and Φ is thus strongly suggested in this context. It is a good thing to have an intuitive feeling for Fourier transforms in this as in other branches of physics.

2.2. Stationary states

We now attempt to find solutions to the Schrödinger equation (2.4)

$$i\hbar\frac{\partial\Psi}{\partial t} = \mathcal{H}\Psi \tag{2.15}$$

under certain restricting conditions.

If $\mathcal{H}$ does not depend explicitly on the time we may try a separation of variables:

$$\Psi(x, y, z, t) = \psi(x, y, z) T(t). \tag{2.16}$$

$\mathcal{H}$ does not operate on T nor $\partial/\partial t$ on ψ. So dividing eq. (2.15) by ψT we obtain:

$$\frac{i\hbar}{T}\frac{dT}{dt} = \frac{1}{\psi}\mathcal{H}\psi. \tag{2.17}$$

It follows that

$$\frac{i\hbar}{T}\frac{dT}{dt} = E', \tag{2.18}$$

$$\frac{1}{\psi}\mathcal{H}\psi = E', \tag{2.19}$$

where E' is a separation constant, independent of t and of x, y, z, and having the dimensions of an energy.

Equation (2.18) has the solution

$$T = T_0\,e^{-iE't/\hbar}. \tag{2.20}$$

Absorbing the constant T_0 into ψ we obtain

$$\Psi = \psi(x, y, z)\,e^{-iE't/\hbar} \tag{2.21}$$

and

$$\mathcal{H}\psi = E'\psi. \tag{2.22}$$

The probability density is

$$\Psi^*\Psi = e^{iE't/\hbar}\,\psi^*\,e^{-iE't/\hbar}\,\psi = \psi^*\psi \tag{2.23}$$

which is independent of time. Hence, without having specified exactly what system we are talking about, we see how the idea of *stationary states* of a system arises, an idea which is implied by the assumption that the Hamiltonian does not depend explicitly on the time. If $\mathscr{H}$ is to describe the stationary states of an atom, we are assuming in particular that the interaction of the atom with a (time-dependent) radiation field is so weak that the presence of the field can be neglected. The atom can then be treated as an isolated system characterized by the value of E', which turns out to be the total energy of the atom. The value of E' has to be found from eq. (2.22) which is a time-independent eigenvalue equation (Schrödinger's time-independent equation). E' may take discrete or continuous eigenvalues.

If the interaction with the radiation field is not neglected a description of this interaction has to be incorporated in the Hamiltonian (this can actually be done via a time-dependent vector potential for the field). Then the atom can no longer be considered as isolated and the concept of stationary states breaks down, except as a zeroth approximation in which it is considered that transitions between stationary states take place with the emission or absorption of electromagnetic radiation. This matter is considered in chapter 3.

2.3. Expectation values

We assume that with every observable (that is to say, a measurable dynamical quantity such as position, momentum, angular momentum, energy, etc.) we can associate an operator which operates on a wave function. For such an operator **A** we can define an *expectation value* $\langle\mathbf{A}\rangle$ by the equation

$$\langle\mathbf{A}\rangle = \int \Psi^*\mathbf{A}\Psi \, d\tau \tag{2.24}$$

subject to the normalization condition

$$\int \Psi^*\Psi \, d\tau = 1 \tag{2.25}$$

where the integration is taken over all space (it is necessary therefore that $\Psi^*\Psi$ be integrable over all space). $\langle\mathbf{A}\rangle$ has the nature of an average value of the results of many measurements of the observable **A** when the measurements are made on the system in the state Ψ. If u is an eigenfunction of **A** with eigenvalue a, that is, if

$$\mathbf{A}u = au \tag{2.26}$$

then

$$\langle \mathbf{A} \rangle = \int u^* \mathbf{A} u \, d\tau = a \int u^* u \, d\tau \qquad (2.27)$$

or

$$\langle \mathbf{A} \rangle = a \qquad (2.28)$$

where u is normalized. When the system is in the state u one will always obtain the result a for a measurement of $\mathbf{A}$ (see eq. (A.3) of appendix A).

We now confirm that the stationary states of section 2.2 are characterized by fixed values of the energy in the sense that $\langle \mathcal{H} \rangle$, the expectation of the energy operator, is the value E of the energy one would obtain as the result of a measurement when the system is in a stationary state; for we have

$$\langle \mathcal{H} \rangle \equiv E = \int \Psi^* i\hbar \frac{\partial \Psi}{\partial t} \, d\tau$$

$$= \int e^{iE't/\hbar} \psi^* (i\hbar) (-iE'/\hbar) \psi \, e^{-iE't/\hbar} \, d\tau$$

$$= E' \int \psi^* \psi \, d\tau$$

$$= E' \quad (\psi \text{ normalized}) \qquad (2.29)$$

which identifies E' for stationary states with $\langle \mathcal{H} \rangle$.

We are now in a position to discuss the gross structure of the hydrogen atom by the method of wave mechanics, that is, by the solution of Schrödinger's equation (2.22) as a differential equation with boundary conditions. We do this in the next section. Later on we shall make use of some more general theorems of quantum mechanics. A summary of the theorems we shall need is given in appendix A.

2.4. Solution of Schrödinger's equation for a Coulomb field

The non-classical aspect of this discussion of Schrödinger's equation lies in the introduction of the wave function. We have seen that the recipe, eq. (2.2), for converting a classical momentum to its operator form depends on the choice of the co-ordinate representation $\Psi(x, y, z, t)$ rather than, say, the momentum representation $\Phi(p_x, p_y, p_z, t)$. But so far the wave function has only been a device, with the formal interpretation of a probability amplitude, on which operators representing dynamical variables can operate and thus throw the equation of motion of a system into the form of a wave equation. To get a feeling for the actual dependence of the wave function on the co-ordinates in a physical problem we must solve a differential equation. The simplest problem is that of the hydrogen atom, and we shall see what form the wave function takes.

The hydrogen atom is assumed to consist of a point nucleus of charge Ze, mass M, and an electron of charge $-e$, mass m_0. The only term in the potential energy which we consider here is the electrostatic interaction $V(r) = -Ze^2/4\pi\varepsilon_0 r$ between these two charges. $V(r)$ depends only on the scalar distance r of the electron from the nucleus. This is the key to the whole problem, and we can deduce certain results from the mere fact that $V(r)$ represents a *central field* without knowing that the form of this field is a Coulomb one. $V(r)$ represents a central field in the relative co-ordinate r. By a separation of variables we can eliminate the centre-of-mass motion of the atom (see problem 2.1) and work in spherical polar co-ordinates (r, θ, ϕ). Schrödinger's time-independent equation for the relative motion has the following simple form if the reduced mass of the electron–nuclear system, $m = m_0 M/(m_0 + M)$, is introduced:

$$\left(\frac{-\hbar^2}{2m}\nabla^2 + V(r)\right)\psi = E\psi, \qquad (2.30)$$

$$V(r) = -Ze^2/4\pi\varepsilon_0 r \qquad (2.31)$$

where

$$\nabla^2 = \frac{1}{r^2}\left\{\frac{\partial}{\partial r}\left(r^2 \frac{\partial}{\partial r}\right) + \frac{1}{\sin\theta}\frac{\partial}{\partial\theta}\left(\sin\theta\frac{\partial}{\partial\theta}\right) + \frac{1}{\sin^2\theta}\frac{\partial^2}{\partial\phi^2}\right\}, \quad (2.32)$$

and the origin of co-ordinates is taken at the nucleus.

The most important result which follows from the fact that $V(r)$ is a central field is that we can achieve a separation in radial and angular parts:

$$\psi(r, \theta, \phi) = R(r)Y(\theta, \phi). \qquad (2.33)$$

A further separation is possible:

$$Y(\theta, \phi) = \Theta(\theta)\Phi(\phi). \qquad (2.34)$$

These separations lead to three differential equations, one each in the co-ordinates r, θ, ϕ (see problem 2.2):

$$\frac{d^2\Phi}{d\phi^2} = -m_l^2\Phi, \qquad (2.35)$$

$$-\frac{1}{\sin\theta}\frac{d}{d\theta}\left(\sin\theta\frac{d\Theta}{d\theta}\right) + \frac{m_l^2}{\sin^2\theta}\Theta = l(l+1)\Theta, \qquad (2.36)$$

$$\frac{1}{r^2}\frac{d}{dr}\left(r^2\frac{dR}{dr}\right) - \frac{l(l+1)}{r^2}R - \frac{2m}{\hbar^2}V(r)R + \frac{2mE}{\hbar^2}R = 0, \quad (2.37)$$

where $-m_l^2$ and $l(l+1)$ are separation constants. These constants are written in this way because we shall soon attach significance to them.

2.4. Solution of Schrödinger's equation for a Coulomb field

The combination of eqs. (2.35) and (2.36) gives the equation for the total angular function $Y(\theta, \phi)$:

$$\left\{ -\frac{1}{\sin\theta} \frac{\partial}{\partial\theta} \left(\sin\theta \frac{\partial}{\partial\theta} \right) - \frac{1}{\sin^2\theta} \frac{\partial^2}{\partial\phi^2} \right\} Y = l(l+1)Y. \quad (2.38)$$

Equations (2.35) and (2.36) are both eigenvalue equations. Purely mathematical considerations† show that the only acceptable solutions to eq. (2.36) are those for which Θ is differentiable and behaves regularly for $\cos\theta = \pm1$. The solutions are proportional to the associated Legendre functions $P_l^{m_l}(\cos\theta)$ which are polynomials of degree l where

$$l = 0, 1, 2, \cdots \quad (2.39)$$

and m_l is also an integer, positive or negative, restricted to the values

$$|m_l| \leqslant l. \quad (2.40)$$

The solution of eq. (2.35) is

$$\Phi = e^{im_l\phi} \quad (2.41)$$

where, from the preceding discussion, m_l is restricted to certain integral values. Thus Φ, like Θ, is a single-valued function of its argument. This single-valuedness is not a necessary postulate of wave-mechanics, but a result derived from the condition that we are working in a central field in (r, θ, ϕ) space.

The proportionality constant in $\Theta = N_l^{m_l} P_l^{m_l}(\cos\theta)$ may be chosen so that the total angular function Y is normalized to unity. Then Y, to which are attached the labels l, m_l, is a *spherical harmonic*

$$Y_l^{m_l}(\theta, \phi) = (-1)^{m_l} \left[\frac{(2l+1)}{4\pi} \frac{(l-m_l)!}{(l+m_l)!} \right]^{1/2} P_l^{m_l}(\cos\theta)\, e^{im_l\phi}$$

$$\text{for } m_l \geqslant 0 \quad (2.42)$$

and

$$Y_l^{-m_l} = (-1)^{m_l} Y_l^{m_l*},$$

where

$$\int Y_l^{m_l*} Y_l^{m_l}\, d\Omega = \int_0^{2\pi} d\phi \int_0^\pi \sin\theta\, d\theta\, Y_l^{m_l*} Y_l^{m_l} = 1. \quad (2.43)$$

The spherical harmonics have the *orthogonality* property

$$\int Y_{l'}^{m_l'*} Y_l^{m_l}\, d\Omega = \delta_{m_l, m_l'} \delta_{l, l'}. \quad (2.44)$$

† For mathematical solutions to the equations a book on quantum mechanics should be consulted.

Thus we find we have introduced mathematically two integers l and m_l for which we must find a physical interpretation later. A list of some of the spherical harmonics, together with the form of $P_l^{m_l}(\cos\theta)$, is given in table 2.1.

Table 2.1. Spherical harmonics

$$Y_0^0 = \sqrt{\frac{1}{4\pi}}$$

$$Y_1^0 = \sqrt{\frac{3}{4\pi}}\cos\theta$$

$$Y_1^{\pm 1} = \mp\sqrt{\frac{3}{8\pi}}\sin\theta\, e^{\pm i\phi}$$

$$Y_2^0 = \sqrt{\frac{5}{16\pi}}(3\cos^2\theta - 1)$$

$$Y_2^{\pm 1} = \mp\sqrt{\frac{15}{8\pi}}\sin\theta\cos\theta\, e^{\pm i\phi}$$

$$Y_2^{\pm 2} = \sqrt{\frac{15}{32\pi}}\sin^2\theta\, e^{\pm 2i\phi}$$

$$Y_l^{m_l}(\theta,\phi) = (-1)^{m_l}\left\{\frac{(2l+1)}{4\pi}\frac{(l-m_l)!}{(l+m_l)!}\right\}^{1/2} P_l^{m_l}(\cos\theta)\, e^{im_l\phi}, \quad m_l \geqslant 0;$$

$$Y_l^{-m_l} = (-1)^{m_l}\, Y_l^{m_l *};$$

$$P_l^{m_l}(x) = (1-x^2)^{|m_l|/2}\frac{d^{|m_l|}}{dx^{|m_l|}}P_l(x);$$

$$P_l(x) = \frac{1}{2^l l!}\frac{d^l}{dx^l}(x^2-1)^l.$$

$$\text{Normalization:}\quad \int_0^{2\pi}d\phi\int_0^{\pi}\sin\theta\, d\theta\, Y_l^{m_l *}Y_l^{m_l} = 1$$

We repeat that this discussion of the angular part of the wave function has depended on the existence of a central field $V(r)$ but not on the detailed form of $V(r)$.

To obtain the energy eigenvalues E we need to solve eq. (2.37). For this purpose we must now use the explicit form of $V(r) = -Ze^2/4\pi\varepsilon_0 r$. We restrict ourselves to negative values of E which correspond to bound states of the atom and we ensure this by the substitution

$$a^2 = -\frac{\hbar^2}{8mE}. \tag{2.45}$$

Further, we change the variable:

$$\rho = r/a, \tag{2.46}$$

and write

$$\lambda = 2mZe^2 a/4\pi\varepsilon_0\hbar^2. \tag{2.47}$$

2.4. Solution of Schrödinger's equation for a Coulomb field

Equation (2.37) then becomes

$$\frac{1}{\rho^2}\frac{\mathrm{d}}{\mathrm{d}\rho}\left(\rho^2\frac{\mathrm{d}R}{\mathrm{d}\rho}\right) + \left\{\frac{\lambda}{\rho} - \frac{1}{4} - \frac{l(l+1)}{\rho^2}\right\} R = 0. \qquad (2.48)$$

Often a new form of the wave function is chosen:

$$P(r) = rR \qquad (2.49)$$

in which case eq. (2.48) becomes

$$\frac{\mathrm{d}^2 P(r)}{\mathrm{d}\rho^2} + \left\{\frac{\lambda}{\rho} - \frac{1}{4} - \frac{l(l+1)}{\rho^2}\right\} P(r) = 0. \qquad (2.50)$$

The solution of eq. (2.48) for which R remains finite as $r \to \infty$ and which behaves properly at the origin, that is $rR \to 0$ as $r \to 0$, is written

$$R(\rho) = \mathrm{e}^{-\rho/2}\, \rho^l F(\rho) \qquad (2.51)$$

where $F(\rho)$ is a polynomial which is related to the associated Laguerre polynomial. Equation (2.51) is written in this way to make explicit the asymptotic forms $\mathrm{e}^{-\rho/2}$ as $\rho \to \infty$, and ρ^l as $\rho \to 0$. The condition that this power series in ρ should terminate is that λ is an integer,

$$\lambda = n \qquad (n \geqslant l + 1) \qquad (2.52)$$

or

$$n = l + 1 + n' \quad (n' = 0, 1, 2, \ldots). \qquad (2.53)$$

Thus again we have introduced an integer mathematically.

The function R is labelled by two indices n and l. $R_{nl}(r)$ is normalized according to

$$\int_0^\infty R_{nl}^*(r)R_{nl}(r)r^2\, \mathrm{d}r = 1 \qquad (2.54)$$

or

$$\int_0^\infty P_{nl}^*(r)P_{nl}(r)\, \mathrm{d}r = 1, \qquad (2.55)$$

so that the total wave function $\psi_{n,\,l,\,m_l} = R_{nl}(r)\, Y_l^m(\theta,\ \phi)$ is also normalized. A tabulation of R_{nl} for small values of n and l is given in table 2.2. Graphs of R_{nl} are plotted in Fig. 2.1. From these graphs it can be seen that n is a measure of the radial extent of the wave function.

It is clear that there is an important qualitative difference between the wave functions for $l = 0$ and $l \neq 0$. For $l \neq 0$, $R_{nl}\ (r = 0)$ vanishes. Equation (2.50) is in the form of a one-dimensional equation of motion in which the term $-l(l+1)/\rho^2$ describes an effective centrifugal potential which keeps the electron away from the nucleus. On the other hand, for $l = 0$ there is no centrifugal potential term and the probability density at the origin is

$$|\psi_{n00}(0)|^2 = \frac{Z^3}{\pi a_0^3 n^3}.$$ (2.56)

This behaviour is particularly important in the theory of hyperfine structure.

Table 2.2

$$R_{nl}(r)$$

$$R_{10} = \left(\frac{Z}{a_0}\right)^{3/2} 2 \exp\left(-\frac{Zr}{a_0}\right)$$

$$R_{20} = \left(\frac{Z}{2a_0}\right)^{3/2} 2 \left(1 - \frac{Zr}{2a_0}\right) \exp\left(-\frac{Zr}{2a_0}\right)$$

$$R_{21} = \left(\frac{Z}{2a_0}\right)^{3/2} \frac{2}{\sqrt{3}} \left(\frac{Zr}{2a_0}\right) \exp\left(\frac{-Zr}{2a_0}\right)$$

$$R_{30} = \left(\frac{Z}{3a_0}\right)^{3/2} 2 \left[1 - 2\frac{Zr}{3a_0} + \frac{2}{3}\left(\frac{Zr}{3a_0}\right)^2\right] \exp\left(\frac{-Zr}{3a_0}\right)$$

$$R_{31} = \left(\frac{Z}{3a_0}\right)^{3/2} \frac{4\sqrt{2}}{3} \left(\frac{Zr}{3a_0}\right) \left(1 - \frac{1}{2}\frac{Zr}{3a_0}\right) \exp\left(\frac{-Zr}{3a_0}\right)$$

$$R_{32} = \left(\frac{Z}{3a_0}\right)^{3/2} \frac{2\sqrt{2}}{3\sqrt{5}} \left(\frac{Zr}{3a_0}\right)^2 \exp\left(\frac{-Zr}{3a_0}\right)$$

$$a_0 = \frac{4\pi\varepsilon_0\hbar^2}{m_0 e^2}$$

Normalization: $\int_0^\infty R_{nl}^* R_{nl} r^2 \, dr = 1$

The negative energy eigenvalues are given by the condition (2.52) and eq. (2.47) which become

$$E_n = -\frac{1}{(4\pi\varepsilon_0)^2} \frac{Z^2 e^4 m}{2\hbar^2 n^2}$$ (2.57)

Equation (2.57) is just the quantized energy found by Bohr. The counting of the *principal quantum number n* matches Bohr's counting: $n = 1, 2, 3, \ldots$ With this choice $n-l-1$ or, from eq. (2.53), n' is just the number of nodes in R_{nl}. (An analysis of the spherical harmonics shows that they have $l - |m_l|$ nodes in the range $0 < \theta < \pi$.) Thus we have an interpretation of the principal quantum number n. It is intimately related to the radial part of the wave function in the central-field approximation.

Equation (2.57) contains the reduced mass m of the system. The equation therefore applies to all the hydrogenic atoms hydrogen, deuterium, tritium, muonium, positronium, etc. The Z-dependence is included, so

2.4. Solution of Schrödinger's equation for a Coulomb field

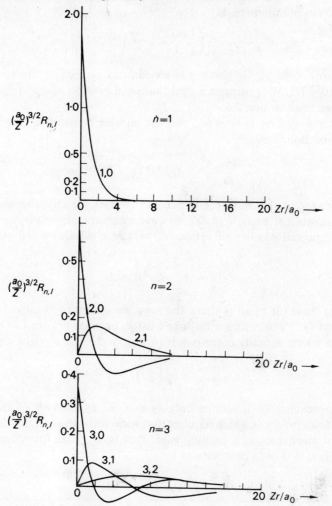

Fig. 2.1. Radial wave functions for hydrogen. Note the different ordinate scales.

the formula applies to He^+, Li^{++}... as well (see also problem 2.9). If we make the isotope dependence explicit we can write

$$E_n = -\frac{1}{(4\pi\varepsilon_0)^2}\left(\frac{1}{2}\frac{e^4 m_0}{\hbar^2}\right)\frac{Z^2}{n^2}\cdot\frac{M}{m_0+M}, \tag{2.58}$$

or

$$E_n = -hcR_\infty \frac{Z^2}{n^2}\frac{M}{m_0+M} \tag{2.59}$$

where R_∞ is a universal constant, the Rydberg, corresponding to the case

21

of a nucleus of infinite mass:

$$R_\infty = \frac{1}{hc}\frac{1}{(4\pi\varepsilon_0)^2}\left(\frac{1}{2}\frac{m_0 e^4}{\hbar^2}\right) = 109,737.31 \text{ cm}^{-1} \qquad (2.60)$$

In electron volts R_∞ is about 13·6 eV, and in cycles per second about $3·3 \times 10^{15}$ Hz. We postpone a quantitative discussion of the *spectrum* of hydrogen until section 2.6.

With $n = 1$, $Z = 1$, $m = m_0$, we have another constant a_0, the radius of the first Bohr orbit:

$$a_0 = \frac{4\pi\varepsilon_0 \hbar^2}{m_0 e^2} = 0·529171 \times 10^{-10}\text{m}. \qquad (2.61)$$

In Bohr's theory the 'radius of an electron orbit' is, for a hydrogenic atom of infinite nuclear mass, $n^2(a_0/Z)$. In wave-mechanics, however, we speak of the expectation value of various powers of r, which we can evaluate from

$$\langle r^k \rangle = \int_0^\infty R_{nl}^* r^k R_{nl} r^2 \mathrm{d}r. \qquad (2.62)$$

A list of these for small positive and negative k is given in table 2.3. We note that $\langle r^{-1} \rangle$ coincides with Bohr's value, but that $\langle r \rangle$ for $l = n - 1$, the case which actually corresponds to Bohr's circular electron orbit, is

$$\langle r \rangle_{l=n-1} = n^2\left(1 + \frac{1}{2n}\right)\frac{a_0}{Z}.$$

This approaches Bohr's value only as $n \to \infty$, an example of the *correspondence principle* which requires that wave mechanics should include classical mechanics as a limiting case: that is, the case for which conceptually $h \to 0$ or in practice $n \to \infty$.

Table 2.3

$$\langle r^k \rangle = \int_0^\infty R_{nl}^* r^k R_{nl} r^2 \mathrm{d}r$$

$$\langle r \rangle = \tfrac{1}{2}[3n^2 - l(l+1)]\frac{a_0}{Z}$$

$$\langle r^2 \rangle = \tfrac{1}{2}[5n^2 + 1 - 3l(l+1)]n^2\left(\frac{a_0}{Z}\right)^2$$

$$\langle r^{-1} \rangle = \frac{1}{n^2}\left(\frac{Z}{a_0}\right)$$

$$\langle r^{-2} \rangle = \frac{2}{(2l+1)n^3}\left(\frac{Z}{a_0}\right)^2$$

$$\langle r^{-3} \rangle = \frac{1}{l(l+\frac{1}{2})(l+1)n^3}\left(\frac{Z}{a_0}\right)^3$$

Bohr's model of a hydrogen atom in its most classical aspect of plane-tary particle motion is superseded in wave mechanics, except in so far as it contributes to the language (e.g., the word 'orbit') and is interpreted in the spirit of the correspondence principle. The demand for a model as a pictorial representation is satisfied in wave mechanics by graphical plots of the mathematical results: in particular, in the case of hydrogen, by an energy-level diagram which is a one-dimensional plot of eq. (2.57) and by plots of spatial distribution functions such as Fig. 2.1.

Bohr's theory contained the concept of orbital angular momentum. We now discuss this in relation to Schrödinger's equation.

2.5. The quantum numbers *l* and m_l

The classical orbital angular momentum

$$\mathscr{L} = \mathbf{r} \times \mathbf{p} \tag{2.63}$$

becomes the operator

$$\hbar\mathbf{l} = -i\hbar\mathbf{r} \times \nabla, \tag{2.64}$$

where we have introduced the factor $\hbar$ into the definition of $\mathbf{l}$ in order to avoid carrying units of $\hbar$ later. In spherical polar co-ordinates the z-component of $\mathbf{l}$, which from eqs. (2.63) and (2.64) is given by

$$\hbar l_z = xp_y - yp_x = -i\hbar x \frac{\partial}{\partial y} + i\hbar y \frac{\partial}{\partial x}, \tag{2.65}$$

becomes

$$\hbar l_z = -i\hbar \frac{\partial}{\partial \phi}. \tag{2.66}$$

By the application of commutation relations similar to eq. (2.14) we find that

$$\mathbf{l}^2 = \frac{1}{\hbar^2}(\mathbf{r} \times \mathbf{p}) \cdot (\mathbf{r} \times \mathbf{p})$$

becomes (see problem 2.4)

$$\mathbf{l}^2 = -\frac{1}{\sin^2\theta}\left\{\sin\theta \frac{\partial}{\partial\theta}\left(\sin\theta \frac{\partial}{\partial\theta}\right) + \frac{\partial^2}{\partial\phi^2}\right\}. \tag{2.67}$$

Now eq. (2.38) is just the eigenvalue equation

$$\mathbf{l}^2 Y_l^{m_l} = l(l+1)Y_l^{m_l}, \quad l = 0, 1, 2, \ldots \tag{2.68}$$

We also have

$$l_z Y_l^{m_l} = m_l Y_l^{m_l}, \quad m_l = -l, -l+1, \cdots, 0, \cdots, l. \tag{2.69}$$

23

Thus we verify that the spherical harmonics $Y_l^{m_l}$ are eigenfunctions of l^2 and l_z simultaneously. Since l^2 and l_z do not operate on the radial part of the wave function, ψ itself is a simultaneous eigenfunction of l^2, l_z and of $\mathcal{H}$. (The z-component of l is singled out merely under the same convention by which the axis of spherical polar co-ordinates is called the z-axis.) So we have interpreted the quantum numbers m_l and l: m_l is the expectation value of the operator corresponding to the z-component of the classical orbital angular momentum, and l is its maximum value. We can form a picture (the beginnings of another model, the vector model) of a *vector* l whose projection on the z-axis is allowed to take only the values m_l differing from each other by integers, and whose length is labelled by l where the actual eigenvalue of the operator l^2 is $l(l + 1)$. The fact that there are only $2l + 1$ eigenfunctions of l_z for a given l, and not an infinite number, is referred to as *space quantization*. (See Fig. 2.2.)

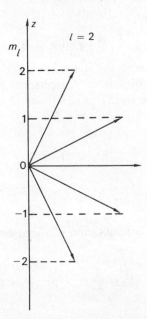

Fig. 2.2 Space quantization, showing the $2l + 1$ projections of an angular momentum vector l on the z-axis for $l = 2$.

An important property of the vector operator l is that its components do not commute with each other. The commutation relations are in fact (see problem 2.5)

$$[l_x, l_y] \equiv l_x l_y - l_y l_x = i l_z \quad \text{and cyclically,} \qquad (2.70)$$

which follow from eqs. (2.65) and (2.14). The three such relations can be

2.5. The quantum numbers l and m_l

combined in the symbolic form

$$\mathbf{l} \times \mathbf{l} = i\mathbf{l}. \tag{2.71}$$

(Remember that this is an operator equation as distinct from the equation for classical vectors: $\mathbf{l} \times \mathbf{l} = 0$. This important difference need not deter us from using the same notation for both quantities.) Although the components of $\mathbf{l}$ do not commute with each other, each component of $\mathbf{l}$ commutes with $\mathbf{l}^2$. Hence we have the situation, discussed in connection with eq. (A.16) of appendix A, in which l_z and $\mathbf{l}^2$ should have simultaneous eigenfunctions. (As already explained l_z is singled out by convention.) We have just found that this is the case: in eqs. (2.68) and (2.69) the $Y_l^{m_l}$ are the simultaneous eigenfunctions of $\mathbf{l}^2$ and l_z.

It is also true that both $\mathbf{l}^2$ and l_z commute with the Hamiltonian of eq. (2.30). The consequence of this is of great general importance. If an operator $\mathbf{B}$ does not depend explicitly on the time, then the time rate of change of its expectation value $\langle \mathbf{B} \rangle = \int \Psi^* \mathbf{B} \Psi \, d\tau$ is

$$\frac{d\langle \mathbf{B} \rangle}{dt} = \int \frac{\partial \Psi^*}{\partial t} \mathbf{B} \Psi \, d\tau + \int \Psi^* \mathbf{B} \frac{\partial \Psi}{\partial t} \, d\tau. \tag{2.72}$$

Now $\mathcal{H} = i\hbar \, (\partial/\partial t)$, so

$$i\hbar \frac{d\langle \mathbf{B} \rangle}{dt} = -\int \left(-i\hbar \frac{\partial \Psi^*}{\partial t} \right) \mathbf{B} \Psi \, d\tau + \int \Psi^* \mathbf{B} \left(i\hbar \frac{\partial \Psi}{\partial t} \right) d\tau$$

$$= -\int \left(\mathcal{H}^* \Psi^* \right) \mathbf{B} \Psi \, d\tau + \int \Psi^* \mathbf{B} \mathcal{H} \Psi \, d\tau. \tag{2.73}$$

But $\mathcal{H}$ is a Hermitian operator (eq. A.6), hence

$$i\hbar \frac{d\langle \mathbf{B} \rangle}{dt} = -\int \Psi^* \mathcal{H} \mathbf{B} \Psi \, d\tau + \int \Psi^* \mathbf{B} \mathcal{H} \Psi \, d\tau$$

$$= \int \Psi^* (\mathbf{B}\mathcal{H} - \mathcal{H}\mathbf{B}) \Psi \, d\tau \tag{2.74}$$

$$= \langle [\mathbf{B}, \mathcal{H}] \rangle.$$

If $[\mathbf{B}, \mathcal{H}] = 0$, $d\langle \mathbf{B} \rangle/dt = 0$ and $\langle \mathbf{B} \rangle$ represents a *constant of the motion*. That is, with $\mathcal{H}$ and t representing conjugate variables, commutation with $\mathcal{H}$ implies independence of t.

Since $\mathbf{l}^2$ and l_z commute with the Hamiltonian of the hydrogen problem they represent constants of the motion. By the correspondence principle one would argue that classically the orbital angular momentum is a constant of the motion because there is no torque upon it, and hence one might expect this to be true in quantum mechanics. In general, to find the constants of the motion we should seek those variables represented by operators which commute with the Hamiltonian.

The spherical harmonics, which are the eigenfunctions of orbital angular momentum, have an important property under inversion of co-ordinates through the origin: if

$$(r, \theta, \phi) \rightarrow (r, \pi - \theta, \phi + \pi)$$

then

$$Y_l^{m_l}(\pi - \theta, \phi + \pi) = (-1)^l Y_l^{m_l}(\theta, \phi). \tag{2.75}$$

The new sign of $Y_l^{m_l}$ under inversion of co-ordinates depends only on l (see problem 2.6). $Y_l^{m_l}$ is said to have even (odd) *parity* if l is even (odd). Since the radial function $R_{nl}(r)$ is even, ψ itself has even (odd) parity if l is even (odd). We can use an abstract operator P to describe this change of co-ordinates. For any function $\psi(\mathbf{r})$ the definition of P gives

$$P\psi(\mathbf{r}) = \psi(-\mathbf{r}) \tag{2.76}$$

and

$$P^2\psi(\mathbf{r}) = P\psi(-\mathbf{r}) = \psi(\mathbf{r}), \tag{2.77}$$

so P^2 has the eigenvalue $+1$. If $\psi(\mathbf{r})$ is to have a well-defined parity it must be an eigenfunction of P itself: it has either even or odd parity corresponding to the eigenvalues ± 1 of P. Also in general,[†] if there are no external forces acting on a system, the Hamiltonian is invariant under the parity operation, that is P commutes with $\mathscr{H}$. Hence the parity of a wave function is a constant of the motion, and so a wave function of definite parity keeps that parity for all time. We shall return to a discussion of this important symmetry property when we come to treat selection rules for radiative transitions.

2.6. The hydrogen energy spectrum

The energy levels for the gross structure of hydrogen are described by eq. (2.57) with $Z = 1$ and with m equal to the reduced mass of the electron-proton system. Figure 2.3 is an energy-level diagram for hydrogen with n running from 1 for the ground state to ∞ at the ionization limit. Above the ionization limit there is a continuum of energy levels corresponding to the positive energy solutions of Schrödinger's equation with $V = 0$. We have not actually discussed these solutions but they clearly refer to a free electron, whose energy eigenvalues are not quantized.

Spectroscopic observations are concerned with the transitions between the energy levels. We shall see in chapter 3 that selection rules have to do with the symmetry properties of angular momentum. For the principal quantum number n there are no selection rules, and electric dipole radia-

---

[†] Actually, the statements made here are not true for systems involved in the so-called 'weak' interaction, as in beta decay. But in discussing atomic structure we confine ourselves to the electromagnetic interactions for which conservation of parity does hold.

2.6. The hydrogen energy spectrum

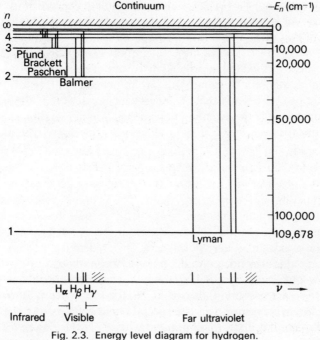

Fig. 2.3. Energy level diagram for hydrogen.

tion, whose frequency obeys the condition

$$hv_{nn'} = E_n - E_{n'}, \qquad (2.78)$$

is emitted in transitions between the stationary states for all n and n'.

The observed spectral lines form simple series the first few of which have been given names: Lyman, Balmer, Paschen, Brackett, Pfund. Each series converges to a limit beyond which a continuous spectrum is observed under favourable excitation conditions in a discharge tube. The first three series are well separated from each other: notice that the excited energy levels all lie in the top quarter of the diagram, so even the lowest-frequency member of the Lyman series lies in the far ultraviolet at 1,216 Å and the series runs to a limit of 912 Å corresponding to the ionization potential of hydrogen: 109,678 cm^{-1} or 13·595 eV.† The Balmer series is the best-known because, unlike the others, it lies in the visible. Its individual members are given names: H_α, H_β, H_γ, etc. The lowest-frequency member, H_α, occurs at 6,563 Å and is responsible for the reddish colour of a clean hydrogen discharge.

The entire spectrum of deuterium is shifted from that of hydrogen because deuterium has a different reduced mass correction. The ratio of the frequencies of deuterium to those of hydrogen is 1·00027. Thus the

† Spectroscopists usually quote wavelengths referred to measurements in dry air at 18°C and 760 mm Hg. Wavenumbers, on the other hand, refer to vacuum conditions. The refractive index of air is about 1·0003.

lines of deuterium are shifted towards the violet. The ionization potential of deuterium is 109,708 cm^{-1}, and the wavelength of D_α is about 1·77 Å less than that of H_α. This is an example of *isotope shift* arising from the *normal mass effect*. Such a large separation can easily be resolved, and indeed this effect led to the original discovery of deuterium.

From a theoretical standpoint an important feature of the gross structure of hydrogen is that the energy eigenvalues (eq. (2.57)) depend only on n whereas the eigenfunctions depend on n, l, and m_l. There are n values of l for each n and $2l + 1$ values of m_l for each l. Therefore there are

$$\sum_0^{n-1} (2l + 1) = n^2$$

wave functions for each energy level. The energy levels are said to be n^2-fold degenerate. The degeneracy with respect to m_l is intuitively obvious, for m_l describes the projection of the vector $\mathbf{l}$ on a co-ordinate axis (the z-axis) in space: if this axis is not defined by some physical criterion but is arbitrarily chosen (as here) for mathematical convenience, then the physically observable energy cannot depend on this arbitrary choice. We are assuming, of course, that x, y, z space is isotropic. The degeneracy with respect to l is a special result arising from the fact that for hydrogen-like atoms (and for these atoms only) $V(r)$ is a pure Coulomb field: $V(r) \propto r^{-1}$; and further, that Coulomb's law, originally derived from observation in classical electrostatics, is valid over the entire range of r of interest in atoms. This is called an *accidental* degeneracy. It occurs likewise in the equivalent classical problem—the Kepler problem of planetary motion under an inverse square law of force, in which all orbits of a given semi-major axis have the same total energy independent of their eccentricity.

We shall find that the degeneracy of a level is *lifted* to a certain extent when we consider further interaction terms which we have so far neglected. For example, when a magnetic or electric field is applied, the field direction defines an axis in space and the energy of the atom comes to depend on m_l. We have already said that we shall use the method of perturbation theory for treating these additional (small) interactions. We shall nearly always find that the perturbation is being applied to an initially degenerate system so that we have to use the so-called degenerate perturbation theory. Unfortunately this is not quite as straightforward as non-degenerate perturbation theory and the slight additional complications have to be faced. A summary of the results of perturbation theory is given for future reference in appendix B.

Before we leave the gross structure of hydrogen it is worth noting that in this special case a rather rare situation occurs: states of *opposite parity* are degenerate with each other for $n > 1$, that is, different states with l both even and odd have the same energy owing to the accidental degeneracy with respect to l. This has interesting consequences for the Stark effect in hydrogen (chapter 8).

Problems

(Those problems marked with an asterisk are more advanced.)

2.1. Set up Schrödinger's time-independent equation for the motion of a proton and an electron under the interaction $V(r)$ where r is the distance between the proton and the electron. By a separation of variables derive eq. (2.30) for the relative motion

$$\left\{ -\frac{\hbar^2}{2m} \nabla^2 + V(r) \right\} \psi = E\psi$$

where m is the reduced mass and ∇^2 is expressed in relative co-ordinates.

2.2. By a separation of variables in spherical polar co-ordinates, $\psi(r, \theta, \phi) = R(r)\Theta(\theta)\Phi(\phi)$, derive from the Schrödinger equation (2.30) three separate equations for R, Θ and Φ, eq. (2.35), (2.36), and (2.37).

2.3. The general form of $R_{n\,;\,l=n-1}$ is

$$\left[(2n)! \right]^{-1/2} \left(\frac{2}{na_0} \right)^{3/2} \left(\frac{2r}{na_0} \right)^{n-1} e^{-r/na_0}.$$

With this function find $\langle r^2 \rangle$ and $\langle r \rangle^2$ and show that

$$\delta r \equiv (\langle r^2 \rangle - \langle r \rangle^2)^{1/2} = (2n + 1)^{-1/2} \langle r \rangle.$$

Attach a classical meaning to this result for large n.

$$\left[\int_0^\infty x^n e^{-x} \, dx = n! \right]$$

*2.4. Show, by using commutation relations, that $\mathbf{l}^2$, which is $(1/\hbar^2)$ $(\mathbf{r} \times \mathbf{p}) \cdot (\mathbf{r} \times \mathbf{p})$ becomes

$$\frac{1}{\hbar^2} \left\{ \mathbf{r}^2 \mathbf{p}^2 - (\mathbf{r} \cdot \mathbf{p})(\mathbf{r} \cdot \mathbf{p}) + i\hbar\, (\mathbf{r} \cdot \mathbf{p}) \right\}.$$

Hence show that, with

$$\nabla^2 = \frac{1}{r^2} \left\{ \frac{\partial}{\partial r} \left(r^2 \frac{\partial}{\partial r} \right) + \frac{1}{\sin\theta} \frac{\partial}{\partial \theta} \left(\sin\theta \frac{\partial}{\partial \theta} \right) + \frac{1}{\sin^2\theta} \frac{\partial^2}{\partial \phi^2} \right\},$$

$$\mathbf{l}^2 = -\frac{1}{\sin^2\theta} \left\{ \sin\theta \frac{\partial}{\partial \theta} \left(\sin\theta \frac{\partial}{\partial \theta} \right) + \frac{\partial^2}{\partial \phi^2} \right\},$$

which is eq. (2.67).

*2.5. From the commutation relations $xp_x - p_x x = i\hbar$ (and cyclically) derive the commutation relations $l_x l_y - l_y l_x = il_z$ (and cyclically) for orbital angular momentum, where $\hbar l_x = yp_z - zp_y$ (and cyclically).

2.6. Verify that for spherical harmonics $Y_l^{m_l}(\theta, \phi)$

$$Y_l^{m_l}(\pi - \theta, \phi + \pi) = (-1)^l\, Y_l^{m_l}(\theta, \phi).$$

(Use the formulae of table 2.1.)

2.7. Show pictorially the dependence of the hydrogenic wave functions on angle by plotting the function $|Y_l^{m_l}(\theta, \phi)|^2$ on a polar diagram for $l = 0, 1$, and 2.

2.8. Show that the Paschen series in the gross structure of hydrogen does not overlap the Balmer series and find which series are the first to overlap.

2.9. A negative muon is a particle with a mean life of about 2×10^{-6} s, which is long compared with times characteristic of radiative transitions in atoms. It interacts with a proton electromagnetically through its charge, but any other interaction between a muon and a nucleon is extremely weak. A negative muon can be captured by an atom, and during the course of its lifetime it can form with the nucleus a hydrogen-like muonic atom. In this respect the muon can be treated like an electron except that its rest mass is 207 times that of the electron.

Consider the muonic transition $1s$—$2p$ (neglecting fine structure effects) in titanium ($Z = 22$, $A = 48$).

(a) What is the radius a_μ/Z of the muonic 'first Bohr orbit' in Ti? Compare this with a_0/Z for an electron.

(b) Hence show that, whereas electronic X-ray levels have to be treated in terms of $Z_{\text{eff}} = Z - \sigma$ where $\sigma(n, l)$ describes screening of the nuclear charge by other electrons, for muonic levels we can ignore screening by the electrons and can take $\sigma = 0$.

(c) What is the radiated energy (in MeV) for the muonic transition $1s$—$2p$ in Ti on the assumption that the nucleus is a point charge?

(d) Evaluate the effective radius R of the Ti nucleus from the formula $R = 1\cdot 2 \times 10^{-13}\, A^{1/3}$ cm.

(e) Show that the potential energy $V(r)$ of a particle of charge $-e$ at radius r inside a charged sphere of radius R in which the charge Ze is uniformly distributed throughout the sphere is

$$V(r) = \frac{-Ze^2}{4\pi\varepsilon_0 R}\left\{\frac{3}{2} - \frac{r^2}{2R^2}\right\}, \quad r \leqslant R$$

and outside,

$$V(r) = -Ze^2/4\pi\varepsilon_0 r, \quad r \geqslant R$$

where $V(\infty) = 0$.

(f) Evaluate the minimum (i.e., most negative) potential energy (in MeV) for a muon or electron inside a Ti nucleus according to the model of problem (e). Compare V_{min} with the rest energy of the muon and of the electron. Hence, justify the use of a non-relativistic approximation *for the muon* in what follows.

(g) Use first-order perturbation theory, with hydrogenic wave functions, as a crude estimate to show that the finite nuclear size leads in this approximation to an energy shift of the muonic $1s$ level of

$$\Delta E = \langle V - V_p \rangle \approx \frac{1}{4\pi\varepsilon_0} \frac{2}{5} \frac{Ze^2}{(a_\mu/Z)} \left(\frac{R}{a_\mu/Z} \right)^2$$

where V is the potential energy of problem (e) and V_p is that for a point nucleus. Use the further approximation that $|\psi(r)|^2 \approx |\psi(0)|^2$ in the range $0 \leqslant r \leqslant R$.

(h) Why is the corresponding shift for the $2p$ level much smaller? We neglect it.

(i) Re-evaluate the radiated energy for the muonic transition $1s - 2p$ in Ti on the assumption that the nucleus is an extended charge with a uniform charge distribution over a sphere of radius R. (The experimental result is about 0·95 MeV.)

(j) Repeat problems (a) and (d) for Pb $(Z = 82; A = 208)$. For Pb the treatment of the other parts of this problem would be quite inadequate.

2.10. Positronium is the bound system of an electron and a positron. Give a quantitative account of the gross structure of this system.

2.11. The radial Schrödinger equation for hydrogen in its one-dimensional form (eq. (2.50)) may be written as

$$\frac{-\hbar^2}{2m} \frac{d^2 P_{nl}}{dr^2} + U_l(r) P_{nl} = E_n P_{nl},$$

where

$$U_l(r) = -\frac{Ze^2}{4\pi\varepsilon_0 r} + \frac{\hbar^2}{2m} \frac{l(l+1)}{r^2}$$

is an effective potential energy.

(a) Make a sketch of $U_l(r)$ as a function of $x = Zr/a_0$ for $l = 1$.

(b) Find an expression in terms of n and $l (\neq 0)$ for those values of x for which $E_n = U_l(r)$. These values, x_1 and x_2, are called the classical turning points. A classical particle would be confined to the region between these points, where the kinetic energy is positive. Evaluate x_1 and x_2 for $n = 3$, $l = 1$.

(c) Show that within the range $x_1 < x < x_2$ the wave function $P_{nl}(x)$ is concave towards the x-axis, but outside it $P_{nl}(x)$ is convex towards the x-axis.

(d) How many nodes does the function P_{31} have? Show that all the nodes of P_{nl} must lie within the classical region $x_1 < x < x_2$.

(e) Make a qualitative sketch of the wave function $P_{31}(x)$ based on this analysis.

(f) Repeat the problem for $n = 3$, $l = 0$ and discuss the differences between the cases of $l = 0$ and $l \neq 0$.

3. Radiative transitions

In this chapter we want to discuss the interaction of an atom with an electromagnetic radiation field. A rigorous treatment of this problem has to overcome many difficulties and we shall not attempt to reproduce such a treatment. Our aim will be to find the selection rules for the atom and to indicate how relative intensities of spectral lines may be calculated.

In speaking of an atom in this context we imply that we can in fact think of the atom as an entity separate from the radiation field, that is, we assume the approximation in which the atom and the field are loosely coupled. In this approximation the wave function describing the whole system, atom plus field, is separable into a product of functions, one for the atom and one for the field, each describing one system in the absence of the other. In this way we retain the concept of stationary states of an atom as an approximation. We shall confine the discussion to the discrete energy states of an atom.

3.1. Einstein's *A* and *B* coefficients

It is useful first of all to consider Einstein's treatment† of the interaction of radiation with matter. For simplicity we consider at first just two energy levels E_1 and E_2 of an atom, between which transitions are possible with absorption or emission of radiation of circular frequency ω where

$$\hbar\omega = E_2 - E_1 \quad (E_2 > E_1); \tag{3.1}$$

(see Fig. 3.1). We assume that these energy levels are g_1-fold and g_2-fold degenerate, and that in an assembly of such atoms the populations of levels 1 and 2 at time t are N_1 and N_2 per unit volume. Then three radiation processes are postulated: *spontaneous emission*, for which the rate of change of N_2 depends on the population N_2 in the upper level; *absorption*, for which the rate of change of N_2 depends on the population N_1 in the lower level and also on the energy density per unit frequency range of the radiation $\rho(\omega)$ where ω satisfies eq. (3.1); and *induced emission*, for which the rate of change of N_2 depends on N_2 and $\rho(\omega)$. We can formulate an

† A. Einstein, *Physikalische Zeitschrift* **18**, 121, 1917. This paper has been translated and reprinted in a book by D. ter Haar: *The Old Quantum Theory*, Pergamon Press, 1967.

3.1. Einstein's A and B coefficients

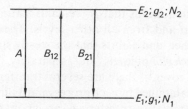

Fig. 3.1. Emission and absorption processes.

expression for the rate of change of N_2 in terms of Einstein's three probability coefficients A, B_{12}, and B_{21} for spontaneous emission, absorption, and induced emission:

$$-\frac{dN_2}{dt} = \frac{dN_1}{dt} = AN_2 - B_{12}\rho(\omega)N_1 + B_{21}\rho(\omega)N_2, \qquad (3.2)$$

in which the first equality simply conserves the total number of atoms per unit volume, $N_1 + N_2$. The coefficients A, B_{12}, and B_{21}, are assumed to be independent of the populations. (Note that A and the B's do not have the same dimensions.) In the absence of a radiation field, $\rho = 0$ and eq. (3.2) has a solution

$$N_2(t) = N_2(0)\,e^{-At} \qquad (3.3)$$

from which we see that $A = 1/\tau$, the reciprocal of a lifetime τ against spontaneous decay by emission of radiation.

If we define probabilities for emission and absorption per atom per unit time, P_{em} and P_{ab}, by

$$-\frac{dN_2}{dt} = P_{em}N_2 - P_{ab}N_1, \qquad (3.4)$$

we have

$$P_{em} = A + B_{21}\rho(\omega), \qquad (3.5)$$

and

$$P_{ab} = B_{12}\rho(\omega). \qquad (3.6)$$

In eq. (3.5) the two emission processes have been grouped together, and eq. (3.6) describes the *inverse* process, absorption. In the steady state, defined by $dN_2/dt = 0$, eq. (3.4) becomes

$$\frac{P_{ab}}{P_{em}} = \frac{N_2}{N_1} = \frac{B_{12}\rho(\omega)}{A + B_{21}\rho(\omega)} \qquad (3.7)$$

In the more general case in which levels 1 and 2 are just two out of many levels, the condition $dN_2/dt = 0$ still defines the steady state for level 2,

33

but now the condition means that N_2 remains stationary with respect to the total transfer to and from all other levels. The *principle of detailed balancing* goes further, and asserts that the steady state is maintained by transfer between *each pair* of levels i and j separately, in particular between levels 1 and 2. Moreover, if there are several transfer processes involved, the steady state is maintained by each process and its inverse independently, and in considering radiative transfer we can ignore other processes such as atomic collisions. So eq. (3.7) is true even in the more general case, and we shall write, for the steady state,

$$\frac{P_{ij}}{P_{ji}} = \frac{N_j}{N_i} = \frac{B_{ij}\rho(\omega_{ij})}{A_{ji} + B_{ji}\rho(\omega_{ij})} \tag{3.8}$$

where P_{ij} refers to the probability of absorption from the lower level i to the upper level j, P_{ji} refers to the probability of emission, and $\omega_{ij} = \omega_{ji} = (E_j - E_i)/\hbar$. In the absence of a radiation field we can generalize eq. (3.3) at once by writing

$$-\frac{dN_j}{dt} = \sum_i A_{ji}N_j \tag{3.9}$$

whence

$$N_j(t) = N_j(0) \exp\left[-\left(\sum_i A_{ji}\right)t\right] \tag{3.10}$$

and the lifetime of the level j against spontaneous emission to all lower levels i is

$$\tau_j = 1/\left(\sum_i A_{ji}\right). \tag{3.11}$$

We now want to know the ratios $A_{ji} : B_{ij} : B_{ji}$. For the purpose of this evaluation we appeal to the special case in which the atoms are in a steady state of thermal equilibrium at temperature T. Then

$$\frac{N_j}{N_i} = \frac{g_j}{g_i} e^{-\hbar\omega_{ij}/kT}. \tag{3.12}$$

Further, we assume that the radiation is in thermal equilibrium with the atoms in which case $\rho(\omega_{ij})$ has the frequency and temperature dependence given by Planck's equation for black-body radiation at temperature T:

$$\rho(\omega_{ij}) = \frac{\omega_{ij}^2}{\pi^2 c^3} \hbar\omega_{ij} \frac{1}{e^{\hbar\omega_{ij}/kT} - 1}. \tag{3.13}$$

We can now make eqs. (3.8), (3.12), and (3.13) compatible at all temperatures if we take

$$g_i B_{ij} = g_j B_{ji}, \tag{3.14}$$

3.1. Einstein's *A* and *B* coefficients

and

$$A_{ji} = \frac{\omega_{ij}^2}{\pi^2 c^3} \hbar\omega_{ij} B_{ji} = \frac{\omega_{ij}^2}{\pi^2 c^3} \hbar\omega_{ij} \frac{g_i}{g_j} B_{ij}. \tag{3.15}$$

These are the relations which Einstein found. In this treatment we have discussed an *assembly* of atoms, but A_{ji}, B_{ij}, and B_{ji} are parameters associated with the internal structure of one atom, not with the assembly of atoms and not with the radiation field. The argument from thermal equilibrium has been introduced only to find the relation between A and the B's. Once the relationship, eq. (3.15), has been found it can be postulated as generally true for an atom irrespective of whether thermal equilibrium obtains or not.

The radiation is specified here by its density ρ per unit frequency range. We can write an expression for the energy density in a significant way as follows:

$$\rho(\omega) = \frac{\omega^2}{\pi^2 c^3} \bar{n}\hbar\omega. \tag{3.16}$$

The factor $\omega^2/\pi^2 c^3$ is the number of modes per unit volume per unit frequency range (see problem 3.1), $\hbar\omega$ is a quantum of radiation energy of frequency ω, and $\bar{n}$ is the average number of photons per mode. This definition of $\bar{n}$ can be connected with a model of radiation in which the radiation is considered as an assembly of simple harmonic oscillators of energy $\bar{n}\hbar\omega$, where n is a quantum number, so that $\bar{n}$ is the value of n averaged over the assembly of oscillators of frequency ω. (For thermal radiation $\bar{n} = (e^{\hbar\omega/kT} - 1)^{-1}$ and this is the quantity which arises in the derivation of Planck's formula from the statistical mechanics of an assembly of harmonic oscillators.) With eqs. (3.15) and (3.16), eqs. (3.5) and (3.6) can be written, for levels i and j,

$$P_{\text{em}} = \frac{\omega_{ij}^2}{\pi^2 c^3} \hbar\omega_{ij}(1 + \bar{n}_{ij})B_{ji}, \tag{3.17}$$

and

$$P_{\text{ab}} = \frac{\omega_{\cdot j}^2}{\pi^2 c^3} \hbar\omega_{ij} \bar{n}_{ij} \frac{g_j}{g_i} B_{ji}, \tag{3.18}$$

whence

$$\frac{P_{\text{em}}}{P_{\text{ab}}} = \frac{g_i}{g_j} \cdot \frac{1 + \bar{n}_{ij}}{\bar{n}_{ij}}. \tag{3.19}$$

These expressions underline the statement that the total emission process depends on the presence of a classical radiation field through the factor $(1 + \bar{n})$, and that the part of the process attributed to spontaneous emission is connected with the term 1 and persists even when $\bar{n}$ vanishes.

35

Having found how Einstein's A and B are connected we now want to see how B is related to particular properties of an atom.

3.2. Transition probabilities

In the situation in which atom and radiation field are loosely coupled the interaction between them is treated by time-dependent perturbation theory. In the textbooks† on quantum mechanics the problem is treated in two ways: (a) the atomic energy levels are quantized but the radiation field is described classically (semi-classical method); (b) the radiation field also is treated as a quantized system (Dirac method).

The semi-classical method is adequate for induced emission and absorption, but it does not account for spontaneous emission. This is because the classical approximation $\bar{n} \gg 1$ is made, so that in eq. (3.17) spontaneous emission is *assumed* to be negligible right from the start. The approximation corresponds, for thermal radiation, to

$$\bar{n} = (e^{\hbar\omega/kT} - 1)^{-1} \gg 1$$

or

$$\bar{n} \approx kT/\hbar\omega \gg 1, \tag{3.20}$$

which in eq. (3.16) or (3.13) leads at once to the Rayleigh–Jeans approximation for the energy density per unit frequency range:

$$\rho(\omega) \to \frac{\omega^2}{\pi^2 c^3} kT, \quad (kT \gg \hbar\omega). \tag{3.21}$$

Of course, it is well known that in the optical range ($\omega \sim 2\pi \times 10^{15}\,\mathrm{Hz}$; $\hbar\omega/k \sim 5 \times 10^4$ K) spontaneous emission is the dominant process, and so the classical approximation is not a good one. Nevertheless, in the semi-classical method B is found in terms of atomic properties and A is related to it by eq. (3.15). This recipe tells us what we want to know for elementary problems, and turns out to be satisfactory for this purpose because the more rigorous Dirac theory, which treats spontaneous and induced emission together, leads exactly to eq. (3.19), confirming that eq. (3.15) is a correct result. Thus no offence is committed against the principles of quantum electrodynamics in pursuing a semi-classical theory by neglecting 1 with respect to $\bar{n}$. It is just that in doing so one gives up the possibility of understanding a mechanism for spontaneous emission to the extent that quantum electrodynamics gives such an understanding: one can only postulate the existence of spontaneous emission as in the Einstein treatment.

Let us now set up the semi-classical formulation of the interaction of an

† E.g. R. M. Sillitto, *Non-relativistic Quantum Mechanics*, Edinburgh, 1960; J. C. Slater, *Quantum Theory of Atomic Structure*, Vol. I, McGraw-Hill, 1960; J. M. Cassels, *Basic Quantum Mechanics*, McGraw-Hill, 1970.

3.2. Transition probabilities

atom with a classical radiation field to see what further approximations are made. In Schrödinger's equation

$$i\hbar \frac{\partial \Psi}{\partial t} = \mathscr{H} \Psi \tag{3.22}$$

the Hamiltonian for a single-electron atom is modified to take account of the interaction between the electron charge, $-e$, and an applied radiation field. The theory is non-relativistic in the sense that intrinsic spin is omitted.

The applied field is represented in general by scalar and vector potentials ϕ and $\mathbf{A}$ which are related to the electric and magnetic field amplitudes, $\mathbf{E}$ and $\mathbf{B}$, by

$$\mathbf{E} = -\nabla\phi - \frac{\partial \mathbf{A}}{\partial t}, \tag{3.23}$$

$$\mathbf{B} = \mathrm{curl}\ \mathbf{A}. \tag{3.24}$$

The modified Hamiltonian then reads†

$$\mathscr{H} = \frac{1}{2m}(\mathbf{p} + e\mathbf{A})^2 - e\phi + V, \tag{3.25}$$

which we write as

$$\mathscr{H} = \mathscr{H}_0 + \mathscr{H}' \tag{3.26}$$

where

$$\mathscr{H}_0 = \frac{\mathbf{p}^2}{2m} + V \tag{3.27}$$

describes the unperturbed atom and

$$\mathscr{H}' = \frac{e}{2m}(\mathbf{p} \cdot \mathbf{A} + \mathbf{A} \cdot \mathbf{p}) + \frac{e^2}{2m}\mathbf{A}^2 - e\phi \tag{3.28}$$

is treated as a perturbation.

We shall regard the radiation field as an incoherent superposition of plane waves periodic in time for each of which

$$\phi = 0; \qquad \mathbf{A} = \mathbf{A}_0\, e^{i(\omega t - \mathbf{k} \cdot \mathbf{r})} + \mathbf{A}_0\, e^{-i(\omega t - \mathbf{k} \cdot \mathbf{r})}, \tag{3.29}$$

where ω is the circular frequency and $\mathbf{k}$ is the wave vector.

Then from eqs. (3.23) and (3.24) we have

$$\mathbf{E} = 2\omega \mathbf{A}_0 \sin(\omega t - \mathbf{k} \cdot \mathbf{r}), \tag{3.30}$$

$$\mathbf{B} = 2(\mathbf{k} \times \mathbf{A}_0) \sin(\omega t - \mathbf{k} \cdot \mathbf{r}), \tag{3.31}$$

† The introduction of the electromagnetic potentials in this way may come as a shock. For clarification, see, for example, R. H. Dicke and J. P. Wittke, *Introduction to Quantum Mechanics*, Addison-Wesley, 1960; in particular, read the discussion of classical mechanics leading up to their eq. (5.52).

so A_0 is related to the amplitude of the electric field and its direction specifies the polarization of the electric vector. We are also free to choose the gauge

$$\nabla \cdot \mathbf{A} = 0 \qquad (3.32)$$

which ensures that the wave is transverse and also has the convenient property that $\mathbf{p}$ commutes with $\mathbf{A}$ (see problem 3.2), so the perturbation may be rewritten

$$\mathcal{H}' = \frac{e}{m} \mathbf{A} \cdot \mathbf{p} + \frac{e^2}{2m} \mathbf{A}^2. \qquad (3.33)$$

Because of the assumption of a weak interaction between radiation and matter we now neglect the small term quadratic in $\mathbf{A}$ which actually describes processes which involve an interchange of two photons, and we restrict the discussion to the term linear in $\mathbf{A}$ (see problem 3.3). Thus we are dealing with the situation in which a weak external (time-dependent) field induces the atom to emit or absorb one photon. In the language of the quantum number n of a simple harmonic oscillator, we have the selection rule for the field $\Delta n = \pm 1$. Thus finally eq. (3.33) reduces to

$$\mathcal{H}' = \frac{e}{m} \mathbf{A} \cdot \mathbf{p}. \qquad (3.34)$$

The Hamiltonian (3.26) is Hermitian.

In the semi-classical approach the wave function in eq. (3.22) is expanded in terms of the *stationary states of the atom*, since the method is concerned primarily with changes in the state of the atom with time rather than changes in the state of the field:

$$\Psi = \sum_n c_n \psi_n \, e^{-iE_n t/\hbar}. \qquad (3.35)$$

The ψ_n are the space-dependent wave functions satisfying

$$\mathcal{H}_0 \psi_n = E_n \psi_n, \qquad (3.36)$$

in which, for simplicity, we assume no degeneracy, and the c_n are time-dependent normalized coefficients with the following interpretation: given that the atom is in state i at time $t = 0$, i.e., $c_i(0) = 1$, $c_{n \ne i}(0) = 0$, we wish to find the probability $|c_j(t)|^2$ that after a time t the atom is in the state j. Schrödinger's equation becomes a differential equation in the c_n which enables us to find $|c_j(t)|^2$:

$$\sum_n (i\hbar \dot{c}_n + E_n c_n) \psi_n \exp\left(-\frac{iE_n t}{\hbar}\right) = (\mathcal{H}_0 + \mathcal{H}') \sum_n c_n \psi_n \exp\left(-\frac{iE_n t}{\hbar}\right).$$
$$(3.37)$$

3.2. Transition probabilities

Since ψ_n is an eigenfunction of $\mathcal{H}_0$, with eigenvalue E_n, eq. (3.37) becomes

$$\sum_n i\hbar \dot{c}_n \psi_n \exp\left(-\frac{iE_n t}{\hbar}\right) = \sum_n c_n \mathcal{H}' \psi_n \exp\left(-\frac{iE_n t}{\hbar}\right). \qquad (3.38)$$

With the further assumption that $\mathcal{H}'$ is so small that c_n does not change much with time, we use the initial values of c_n on the right-hand side of eq. (3.38) as an approximation:

$$\sum_n i\hbar \dot{c}_n \psi_n \exp\left(-\frac{iE_n t}{\hbar}\right) = \mathcal{H}' \psi_i \exp\left(-\frac{iE_i t}{\hbar}\right) \qquad (3.39)$$

in which $c_i(0) = 1$. Multiplying on the left by ψ_j^* and integrating over spatial co-ordinates, we obtain

$$i\hbar \dot{c}_j = \langle j| \mathcal{H}' |i\rangle \exp\left[\frac{i(E_j - E_i)t}{\hbar}\right]. \qquad (3.40)$$

In order to solve this equation for c_j we must make explicit the time dependence of $\mathcal{H}'$ from eqs. (3.34) and (3.29). For simplicity we shall also consider plane polarized radiation, and introduce a unit vector $\hat{\mathbf{e}}$ to describe the direction of polarization so that $\mathbf{A}_0 = A_0 \hat{\mathbf{e}}$. Writing $(E_j - E_i)/\hbar = \omega_{ji}$, we have

$$i\hbar \dot{c}_j = \langle j| \frac{e}{m} \hat{\mathbf{e}} \cdot \mathbf{p}\, e^{-i\mathbf{k}\cdot\mathbf{r}} |i\rangle A_0\, e^{i(\omega_{ji}+\omega)t}$$

$$+ \langle j| \frac{e}{m} \hat{\mathbf{e}} \cdot \mathbf{p}\, e^{i\mathbf{k}\cdot\mathbf{r}} |i\rangle A_0\, e^{i(\omega_{ji}-\omega)t}, \qquad (3.41)$$

which has the solution, subject to $c_j(0) = 0$,

$$c_j(t) = \langle j| \frac{e}{m} \hat{\mathbf{e}} \cdot \mathbf{p}\, e^{-i\mathbf{k}\cdot\mathbf{r}} |i\rangle A_0 \left\{ \frac{1 - e^{i(\omega_{ji}+\omega)t}}{\hbar(\omega_{ji} + \omega)} \right\}$$

$$+ \langle j| \frac{e}{m} \hat{\mathbf{e}} \cdot \mathbf{p}\, e^{i\mathbf{k}\cdot\mathbf{r}} |i\rangle A_0 \left\{ \frac{1 - e^{i(\omega_{ji}-\omega)t}}{\hbar(\omega_{ji} - \omega)} \right\}. \qquad (3.42)$$

Let us now consider the case of $E_j > E_i$ so that in the transition $i \rightarrow j$ we are dealing with an absorption process. We can ignore the first term in eq. (3.42) in comparison with the second because of its large denominator in the frequency region $\omega \approx \omega_{ji}$ in which absorption will take place. This is called the rotating wave approximation (RWA). It is an excellent approximation in the optical region, and even at radio frequencies it is quite adequate for most purposes. Then

$$|c_j(t)|^2 = |\langle j| \frac{e}{m} \hat{\mathbf{e}} \cdot \mathbf{p}\, e^{i\mathbf{k}\cdot\mathbf{r}} |i\rangle|^2 A_0^2 \frac{\sin^2\{(\omega_{ji} - \omega)/2\}t}{\hbar^2\{(\omega_{ji} - \omega)/2\}^2}. \qquad (3.43)$$

We must now examine the validity† of the way in which we want to apply eq. (3.43). In this section we are trying to find an expression, in terms of atomic properties, for the Einstein B coefficient which is defined by eq. (3.2). That equation is a so-called *rate equation*, meaning that dN_2/dt is independent of time or, in other words, that $|c_j(t)|^2$ is required to be linearly proportional to t. Equation (3.2) is also formulated in terms of a continuous distribution of frequencies, of energy density $\rho(\omega)$ per unit frequency range. (It is the assumption of a continuous distribution of $\omega_{ij} - \omega$ which is important for obtaining a *rate* of transition, and this may come about either because, as here, the frequency ω of the radiation has a continuous range of values or because the difference in energy of the atomic levels, $\hbar\omega_{ij}$, has a spread of values.) The derivation of eq. (3.43) has been carried through for a single frequency so far, and as it stands $|c_j(t)|^2$ increases as t^2 at resonance and oscillates with time away from resonance. For our present application we are not interested in such solutions, and to obtain a constant transition rate $|c_j(t)|^2/t$ we need to deal with the frequency distribution of the incident radiation.

So the next step is to sum over the monochromatic plane waves with which we started, eq. (3.29). Notice that we have delayed doing this until we have reached the stage of discussing probabilities rather than probability amplitudes because we assume that there is no phase relationship between incident waves of different frequencies. The energy per unit volume for each monochromatic wave is proportional to the mean square of the electric field which is expressed in terms of A_0 through eq. (3.30):

$$\tfrac{1}{2}\varepsilon_0 E_0^2 = 2\varepsilon_0\omega^2 A_0^2. \tag{3.44}$$

Summation in eq. (3.43) can therefore be carried out by replacing A_0^2 by $(1/2\varepsilon_0)(\rho(\omega)/\omega^2)\,d\omega$ and integrating between limits ω_1 and ω_2 to cover a range of frequencies which includes the resonant frequency ω_{ji}. Then, dividing by t, we obtain a transition rate

$$|c_j(t)|^2/t = \frac{2}{\varepsilon_0\hbar^2}\,|\langle j|\,\frac{e}{m}\,\hat{\mathbf{e}}\cdot\mathbf{p}\,e^{i\mathbf{k}\cdot\mathbf{r}}\,|i\rangle|^2 \int_{\omega_1}^{\omega_2} \frac{\rho(\omega)}{\omega^2}\frac{\sin^2\{(\omega_{ji}-\omega)/2\}t}{(\omega_{ji}-\omega)^2 t}\,d\omega. \tag{3.45}$$

The integral in eq. (3.45) has the property that in the *limit of large t* it has the value

$$\text{Integral} = \frac{\pi}{2}\frac{\rho(\omega_{ji})}{\omega_{ji}^2},$$

that is, the value of the function $\rho(\omega)/\omega^2$ at the resonance frequency has been picked out. So the transition rate is

$$|c_j(t)|^2/t = \frac{\pi}{\varepsilon_0\hbar^2\omega_{ji}^2}\,|\langle j|\,\frac{e}{m}\,\hat{\mathbf{e}}\cdot\mathbf{p}\,e^{i\mathbf{k}\cdot\mathbf{r}}\,|i\rangle|^2\rho(\omega_{ji}). \tag{3.46}$$

† See R. Loudon, *The Quantum Theory of Light*, chapter 3, OUP, 1973, for a more extensive discussion of this topic.

3.2. Transition probabilities

Having at last obtained this expression for the transition probability per unit time we ought to make several remarks about it.

(a) We have used first-order perturbation theory to derive the result: this implies a low power input. The derivation would not have been valid for an intense laser beam, for example.

(b) We have assumed that the atomic energy levels are infinitely sharp and that the perturbation is applied for a time long compared with the inverse of the bandwidth within which $\rho(\omega)$ does not vary appreciably. If, however, a frequency width γ is attributed to the spectral line because of spontaneous emission from the upper state, the result of eq. (3.46) would only be valid for $\gamma t \ll 1$. The problem arising in many modern experiments in laser spectroscopy, in which the bandwidth of the incident laser radiation is much less than the line width of the transition, would have to be treated by different methods because the above inequalities would not then be satisfied simultaneously.

(c) The rate $|c_j(t)|^2/t$ is just the same as P_{ab} in eq. (3.6) so we can now equate Einstein's coefficient for absorption B_{ij} directly to a factor in eq. (3.46):

$$B_{ij} = \frac{\pi}{\varepsilon_0 \hbar^2 \omega_{ij}^2} \, |\langle j| \frac{e}{m} \hat{\mathbf{e}} \cdot \mathbf{p} \, e^{i\mathbf{k} \cdot \mathbf{r}} |i\rangle|^2. \tag{3.47}$$

(d) For non-degenerate levels, which we have assumed in our derivation, the relation $B_{ij} = B_{ji}$ (the special case of eq. (3.14)) follows at once from the Hermitian property of the perturbation $\mathcal{H}'$. This result is of great importance because it is an example of microscopic reversibility, the quantum-mechanical basis for the principle of detailed balancing which we used at a crucial point in the argument of section 3.1.

(e) Because $\mathcal{H}'$ is linear in $\mathbf{A}$ its time-dependence is of the form $e^{i\omega t}$ (eq. (3.29) represents a particular Fourier component of the radiation field). This time-dependence has led to the result that $|c_j(t)|^2/t$ is only appreciable if energy is conserved in the form of the Bohr frequency condition $\hbar\omega = E_j - E_i$.

(f) We have assumed one-electron atoms in this treatment. To deal with many-electron atoms we should have to sum over electrons, writing $\sum_i \mathbf{p}_i \, e^{i\mathbf{k} \cdot \mathbf{r}_i}$ in eq. (3.46) where $\mathbf{p}_i$ is the momentum of the ith electron and $\mathbf{r}_i$ is the space co-ordinate of the incident wave at the position of the ith electron.

(g) Finally, the dependence of the transition rate on the spatial co-ordinates of the atomic electron is through the square of the matrix element of the perturbation connecting the stationary states i and j:

$$|\langle j| \frac{e}{m} \hat{\mathbf{e}} \cdot \mathbf{p} \, e^{i\mathbf{k} \cdot \mathbf{r}} |i\rangle|^2. \tag{3.48}$$

Let us consider this matrix element at greater length because it is the quantity which determines the selection rules for a transition between the states i and j. The occurrence of this *off-diagonal* matrix element is related directly to the use of stationary states as the basis of the expansion (3.35) of the wave function for this problem. The total Hamiltonian is not diagonal in this representation, so the stationary states become mixed as a result of applying the radiation field, and the degree of admixture is described by the coefficients c_n which are proportional to such off-diagonal matrix elements.

3.3. The electric dipole approximation

We are now in a position to discuss further approximations in the matrix element (3.48). The spatial dependence of the incident wave is through the factor $e^{i\mathbf{k}\cdot\mathbf{r}}$ in which the magnitude of the wave-vector $\mathbf{k}$ is $2\pi/\lambda$. For wavelengths large compared with the size of an atom the exponential may be expanded:

$$e^{i\mathbf{k}\cdot\mathbf{r}} = 1 + i\mathbf{k}\cdot\mathbf{r} + \cdots \tag{3.49}$$

The first term leads to the *electric dipole approximation* of this multipole expansion as we shall show. It is the dominant term in the optical region, for if a_0/Z is taken as the size of an atom

$$\frac{2\pi}{\lambda}\cdot\frac{a_0}{Z} = \frac{a_0}{Zhc}(E_j - E_i) < \frac{a_0}{Zhc}\cdot\frac{Z^2 e^2}{4\pi\varepsilon_0 a_0} = Z\alpha \approx \frac{Z}{137} \tag{3.50}$$

and the approximation is a good one. It means that the amplitude of the wave is approximately constant over the size of the atom. Let us consider the electric vector polarized along the x-axis of the electronic co-ordinate system. Then we need an expression for $\langle j|p_x|i\rangle$. By an extension of eq. (2.74) we can write

$$\langle j|p_x|i\rangle = \langle j|m\dot{x}|i\rangle = \frac{im}{\hbar}\langle j|\mathscr{H}_0 x - x\mathscr{H}_0|i\rangle \tag{3.51}$$

for the unperturbed atom. Since $\mathscr{H}_0$ is Hermitian we have

$$\langle j|\frac{e}{m}p_x|i\rangle = \frac{i}{\hbar}(E_j - E_i)\langle j|ex|i\rangle, \tag{3.52}$$

and the atomic part of the problem is reduced to finding the matrix element between the stationary state i and j of a component of the electric dipole operator for the atom. Thus the transition probability per unit time that an atom has made a transition from state i to state j (both non-degenerate) by absorption of electric dipole radiation of energy density $\rho(\omega_{ji})$ per unit frequency range polarized in the x-direction is

$$P_{\text{ab}} = B_{ij}\rho(\omega_{ji}) = \frac{\pi}{\varepsilon_0\hbar^2}\langle j|ex|i\rangle|^2\rho(\omega_{ji}); \quad x\text{-polarization} \tag{3.53}$$

3.3. The electric dipole approximation

where $\hbar\omega_{ji} = E_j - E_i$, and for unpolarized radiation

$$P_{ab} = \frac{1}{3}\frac{\pi}{\varepsilon_0\hbar^2}|\langle j|e\mathbf{r}|i\rangle|^2\rho(\omega_{ji}); \quad \text{unpolarized} \quad (3.54)$$

where

$$|\langle j|\,e\mathbf{r}\,|i\rangle|^2 = |\langle j|\,ex\,|i\rangle|^2 + |\langle j|\,ey\,|i\rangle|^2 + |\langle j|\,ez\,|i\rangle|^2. \quad (3.55)$$

As pointed out before, the same expression (3.53) or (3.54) applies to induced emission when the levels are non-degenerate. In this semi-classical treatment the probability per unit time for spontaneous emission by electric dipole radiation has to be found from eq. (3.15), with eq. (3.54):

$$A_{ji} = \frac{1}{4\pi\varepsilon_0}\cdot\frac{4}{3}\cdot\frac{\omega_{ji}^3}{\hbar c^3}|\langle j|e\mathbf{r}|i\rangle|^2 \quad \text{unpolarized} \quad (3.56)$$

To find the order of magnitude of A we can put $|\langle j|\,e\mathbf{r}\,|i\rangle|^2 \sim (ea_0)^2$ for a transition allowed by electric dipole radiation, giving

$$A \sim \frac{1}{\hbar c^3}\frac{(ea_0)^2}{4\pi\varepsilon_0}\omega^3, \quad (3.57)$$

from which $A \sim 10^8\ \text{s}^{-1}$ in the optical region. This corresponds to a typical lifetime of an excited state of $\tau \sim 10^{-8}$ s against spontaneous decay by electric dipole radiation. Because of its strong frequency dependence A is negligible at radiofrequencies.

Let us now take up the case of degeneracy. First, suppose the lower level i is g_i-fold degenerate. We shall assume that each non-degenerate state of i can be labelled by specifying an additional quantum number m_i (we use the symbol m because we shall be concerned with degeneracy with respect to orientation in space). Then from the (non-degenerate) level j there are a number of 'channels' for decay to the level i, and

$$A_{ji} = \frac{1}{4\pi\varepsilon_0}\cdot\frac{4}{3}\cdot\frac{\omega_{ji}^3}{\hbar c^3}\sum_{m_i}|\langle j|\,e\mathbf{r}\,|im_i\rangle|^2. \quad (3.58)$$

But if the upper level is also degenerate, with label m_j, the rate of decay from *each* state j, m_j is the same, a result which is intuitively reasonable but which formally depends on the crucial fact, which we quote, that

$$\sum_{m_i}|\langle jm_j|\,e\mathbf{r}\,|\,im_i\rangle|^2 \text{ is independent of } m_j. \quad (3.59)$$

That is, having summed over spatial orientation, we have left no specified axis with which to distinguish physically different m_j. Because of eq. (3.59) we can write

$$A_{ji} = \frac{1}{4\pi\varepsilon_0}\cdot\frac{4}{3}\cdot\frac{\omega_{ji}^3}{\hbar c^3}\sum_{m_i}|\langle jm_j|\,e\mathbf{r}\,|im_i\rangle|^2 \quad (3.60)$$

for both j and i degenerate, and this is the same as eq. (3.58).

Since eq. (3.60) does not have a symmetrical appearance with respect to m_i and m_j, another quantity has been defined. It is the line strength

$$S_{ji} = S_{ij} = \sum_{m_j} \sum_{m_i} |\langle jm_j| \, e\mathbf{r} \, |im_i\rangle|^2 \tag{3.61}$$

which is symmetrical. In view of eq. (3.59) the summation over m_j is merely a counting over the g_j non-degenerate states of j and we therefore have the relation

$$A_{ji} = \frac{1}{4\pi\varepsilon_0} \cdot \frac{4}{3} \cdot \frac{\omega_{ji}^3}{\hbar c^3} \cdot \frac{S_{ji}}{g_j}. \tag{3.62}$$

The total rate of loss of energy from an atom by spontaneous emission of unpolarized electric dipole radiation at the frequency ω_{ji} is

$$-\frac{\mathrm{d}W}{\mathrm{d}t} = A_{ji}\hbar\omega_{ji} = \frac{1}{4\pi\varepsilon_0} \cdot \frac{4}{3} \cdot \frac{\omega_{ji}^4}{c^3} |\langle j| \, e\mathbf{r} \, |i\rangle|^2 \tag{3.63}$$

which depends on ω_{ji}^4. Let us compare this expression with that for a classical electron oscillator obeying, in one dimension, the equation of motion

$$\ddot{x} + \gamma\dot{x} + \omega_0^2 x = 0, \tag{3.64}$$

where ω_0 is the oscillation frequency and $\gamma\dot{x}$ is a damping term representing a weak radiation reaction force. γ has the value

$$\gamma = \frac{1}{4\pi\varepsilon_0} \cdot \frac{2}{3} \cdot \frac{e^2\omega_0^2}{mc^3} \tag{3.65}$$

for an electron of mass m and charge $-e$. That γ is small compared with ω_0 is shown by taking the ratio

$$\gamma/\omega_0 \sim \frac{1}{4\pi\varepsilon_0} \frac{e^2}{mc^3} \omega_0 = \alpha \frac{\hbar\omega_0}{mc^2} \sim \alpha^3 \tag{3.66}$$

where we have identified ω_0 with a radiation frequency typical of atoms in the optical region ($\hbar\omega_0 \sim e^2/4\pi\varepsilon_0 a_0$). In this approximation, therefore, eq. (3.64) has a solution

$$x = x_0 \, \mathrm{e}^{-\gamma t/2} \cos \omega_0 t \tag{3.67}$$

and the energy of the oscillator is

$$W(t) = \tfrac{1}{2}m\dot{x}_{\max}^2 \approx \tfrac{1}{2}m\omega_0^2 x_0^2 \, \mathrm{e}^{-\gamma t} = W(0) \, \mathrm{e}^{-\gamma t}. \tag{3.68}$$

This equation leads to a rate of loss of energy

$$-\frac{\mathrm{d}W}{\mathrm{d}t} = \gamma W \approx \frac{1}{4\pi\varepsilon_0} \cdot \frac{1}{3} \cdot \frac{e^2}{c^3} \omega_0^4 x_0^2, \tag{3.69}$$

3.3. The electric dipole approximation

which is the same as the well-known formula

$$-\frac{dW}{dt} = \frac{1}{4\pi\varepsilon_0} \cdot \frac{2}{3} \cdot \frac{e^2}{c^3} \overline{(\ddot{x})^2} \qquad (3.70)$$

if one takes the mean square acceleration as $\overline{(\ddot{x})^2} \approx \frac{1}{2}x_0^2\omega_0^4$ from eq. (3.67). We generalize eq. (3.69) to include three independent orthogonal oscillators by writing $|\mathbf{r}|^2$ instead of x_0^2, and in comparing this classical formula with eq. (3.63) we find

$$|e\mathbf{r}|^2 = |2\langle j| \, e\mathbf{r} \, |i\rangle|^2. \qquad (3.71)$$

The factor 2 in this comparison between the classical electric dipole moment and the quantum mechanical dipole matrix element comes about because we wrote in eq. (3.67) $x \propto x_0 \cos \omega_0 t$ rather than $x \propto x_0' (e^{i\omega_0 t} + e^{-i\omega_0 t}) = 2x_0' \cos \omega_0 t$. Apart from this small point, the analogy is complete at a single frequency.

But in trying to make a model of a one-electron atom out of three classical oscillators (one for each direction of polarization) we do not take account of the fact that the atom can emit many frequencies ω_{ji} whereas the classical oscillators emit only one frequency. Each one-dimensional oscillator has therefore to be endowed with an 'oscillator strength' $-f_{ji}$ (f_{ji} is defined as a negative number for emission, positive for absorption) to indicate the fraction of the oscillator energy which is emitted into each channel, $j \rightarrow i$. In identifying eq. (3.69) with eq. (3.63)

$$-\frac{dW}{dt} = A_{ji}(\hbar\omega_{ji}) = \gamma W \qquad (3.72)$$

we write

$$A_{ji}\hbar\omega_{ji} = \gamma(-3f_{ji}\hbar\omega_{ji})$$

or

$$A_{ji} = -3\gamma f_{ji}. \qquad (3.73)$$

Equation (3.73) is one way of introducing the definition of the oscillator strength (see problem 3.4). With eq. (3.56) this leads to the expression

$$f_{ji} = -\frac{2}{3} \cdot \frac{m}{e^2} \cdot \frac{1}{\hbar^2} |\langle j| \, e\mathbf{r} \, |i\rangle|^2 \hbar\omega_{ji}. \qquad (3.74)$$

For absorption, the definition of f_{ij} is

$$f_{ij} = -\frac{g_j}{g_i} f_{ji}, \qquad (3.75)$$

so f_{ij} is directly related to Einstein's B for absorption through eq. (3.14) and eq. (3.54):

$$f_{ij} = \frac{2\varepsilon_0}{\pi} \frac{m}{e^2} B_{ij}\hbar\omega_{ij}$$

$$= \frac{2}{3} \cdot \frac{m}{e^2} \cdot \frac{1}{\hbar^2} |\langle i| \, e\mathbf{r} \, |j\rangle|^2 \hbar\omega_{ij}. \qquad (3.76)$$

Finally, in terms of the line strength S_{ij}

$$f_{ij} = \frac{2}{3} \cdot \frac{m}{e^2} \cdot \frac{1}{\hbar^2} \cdot \frac{1}{g_i} S_{ij} \hbar \omega_{ij}. \qquad (3.77)$$

The classical oscillator has been discussed at some length because through it the idea of an oscillator strength has been introduced, and this latter quantity is important because it is the one traditionally used by spectroscopists to describe the relative intensities of spectral lines. From the way it has been introduced as a fraction one suspects that it is subject to a sum rule of the form $\sum_j f_{ij} = 1$ for absorption. This is in fact the case if i is the lowest level in a one-electron spectrum of levels. (For a proof see problem 3.5.) If j is an excited state, emission to lower levels i as well as absorption to higher levels k can take place and the f-sum rule reads

$$\sum_{i,k} (f_{ji} + f_{jk}) = 1 \qquad (3.78)$$

where f_{ji} is a negative fraction. If Z electrons are responsible for the spectrum the sum, more generally, equals Z.

The intensity of a spectral line in emission depends on the population of the excited state as well as on the transition probability. The assembly of atoms in the light source may be in thermal equilibrium, brought about by atomic collisions, in which case a temperature may be ascribed to the assembly and a Boltzmann distribution prevails. A condition often discussed theoretically is that of so-called natural excitation which corresponds to thermal equilibrium at an infinite temperature so that all nondegenerate states are equally populated. Another situation in a light source is that atoms are excited under bombardment by charged particles, mostly electrons, and the relative populations of excited states satisfy a Boltzmann distribution at a so-called *electron temperature*, T_e, which may be of the order of 10,000 K or higher. Thus

$$\frac{N_j}{N_i} = \frac{g_j}{g_i} e^{-(E_j - E_i)/kT_e}. \qquad (3.79)$$

At $T_e = 10^4$ °K, $kT_e \sim 2 \times 10^{14}$ Hz or about 1 eV, so excited atomic states are fairly well populated. In an absorption cell, on the other hand, in which atoms are in thermal equilibrium at, say, room temperature, $T \sim 300$ K and $kT \sim 6 \times 10^{12}$ Hz, so that nearly all the atoms are in the ground state. In absorption spectra transitions from the ground state greatly predominate.

3.4. Selection rules for l and m_l

We now discuss the detailed form of the electric dipole matrix elements for a one-electron atom.

3.4. Selection rules for l and m_l

First of all, it is clear that in a stationary state of well-defined parity the expectation value of the electric dipole moment vanishes:

$$\langle e\mathbf{r} \rangle = e \int \Psi^*_{nlm_l} \mathbf{r}\, \Psi_{nlm_l} \, d\tau = 0 \tag{3.80}$$

because for a reflection of the co-ordinate system through the origin Ψ has the sign of $(-1)^l$, $\Psi^*\Psi$ does not change sign, but $\mathbf{r}$ does. (We ignore for the time being any degeneracy of states of opposite parity, assuming that this degeneracy has been removed.) The vanishing of $\langle e\mathbf{r} \rangle$ for a state of well-defined angular momentum also follows from the principle of invariance under time reversal, but the argument from this principle is more difficult.

The selection rules for m_l and l can be worked out directly with the use of hydrogenic wave functions. If we establish the z-axis as a physically preferred direction, for example by applying a magnetic field along the z-axis (Zeeman effect), thus lifting the degeneracy in m_l, we can treat the rectangular components of $\mathbf{r}$ separately:

(a) $z = r \cos\theta$:

$$\langle n'l'm_l'|z|nlm_l\rangle = N \int_0^\infty R^*_{n'l'} r R_{nl} r^2 \, dr \int_0^\pi P_{l'}^{m_l'*} \cos\theta \, P_l^{m_l} \sin\theta \, d\theta$$

$$\int_0^{2\pi} \exp(im_l'\phi) \exp(-im_l\phi) \, d\phi \tag{3.81}$$

where N is a normalization constant. From the ϕ-dependence

$$\langle n'l'm_l'|z|nlm_l\rangle = 0 \quad \text{unless } m_l' = m_l. \tag{3.82}$$

(b) $x = r \sin\theta \cos\phi$:
omitting the r and θ integrals,

$$\langle n'l'm_l'|x|nlm_l\rangle \propto \int_0^{2\pi} \exp(im_l'\phi) \cos\phi \exp(-im_l\phi) \, d\phi$$

$$= \frac{1}{2} \int_0^{2\pi} \{\exp[i(m_l' - m_l + 1)\phi]$$

$$+ \exp[i(m_l' - m_l - 1)\phi]\} \, d\phi$$

$$= 0 \quad \text{unless } m_l' = m_l \pm 1, \tag{3.83}$$

with a similar expression for the y-component.

Thus we obtain the polarization rules for electric dipole radiation:

$\Delta m_l = 0$ for the electric vector parallel† to the magnetic field (π polarization);

$\Delta m_l = \pm 1$ for the electric vector perpendicular to the magnetic field (σ polarization).

† These statements about the direction of the electric vector can be checked by looking back through eqs. (3.52), (3.42), and (3.30).

The selection rules for l follow from the θ-dependence, but first we are able to state a very strict rule for states of well-defined parity:

$$\Delta l = \text{odd}. \tag{3.84}$$

This is based on the symmetry arguments about the parity of the states. The details of the θ-dependence depend on recursion relations among the $P_l^{m_l}$. For the z-component

$$\cos \theta \, P_l^{m_l} = \frac{(l - m_l + 1)P_{l+1}^{m_l} + (l + m)P_{l-1}^{m_l}}{2l + 1}. \tag{3.85}$$

Since the $P_l^{m_l}$ are orthogonal,

$$\int P_{l'}^{m_l} \cos \theta \, P_l^{m_l} \sin \theta \, d\theta = 0 \quad \text{unless } l' = l \pm 1. \tag{3.86}$$

Similarly, for the x- and y-components we use

$$\sin \theta \, P_l^{m_l-1} = \frac{P_{l+1}^{m_l} - P_{l-1}^{m_l}}{2l + 1} \tag{3.87}$$

to obtain the same rule $l' = l \pm 1$. Thus for electric dipole radiation

$$\Delta l = \pm 1. \tag{3.88}$$

In spontaneous emission the relative intensities of lines which are allowed by the electric dipole selection rules are proportional to

$$N_j \omega^4 \left[\int R_{n'l'} r R_{nl} r^2 \, dr \right]^2 \times F(l, m_l, l', m_l').$$

The function $F(l, m_l, l', m_l')$ arising from the angular part can be worked out once and for all in the case of hydrogen and indeed in the more general case of the central-field approximation for many-electron atoms. The population of the emitting excited state, N_j, depends on the details of excitation in the light source and on cascades in emission from higher states down to the state in question. The radial integral, in general, has to be worked out with a knowledge of radial wave functions, hence the importance of atomic theory applied, for example, to astrophysics. There are no selection rules for n, but clearly the value of the integral depends on the amount of overlap between the radial functions $R_{n'l'}$ and R_{nl}, especially in the outer parts of the atom because the integral is weighted by the factor r. This point is mentioned again in chapter 6 (see problem 6.1).

Relative intensities in the Zeeman effect are discussed in more detail in chapter 8.

3.5. Higher order radiation

In the expansion of $e^{i\mathbf{k} \cdot \mathbf{r}}$, eq. (3.49), we have shown that the first term leads to the electric dipole approximation. Inclusion of the second term,

3.5. Higher order radiation

$i\mathbf{k} \cdot \mathbf{r}$, in this multipole expansion leads to a description of magnetic dipole and electric quadrupole radiation, which we shall now briefly discuss. This term brings in the effect of retardation.

In the matrix element of eq. (3.48) the electron momentum vector $\mathbf{p}$ is projected on to the direction of the amplitude of the vector potential $\mathbf{A}_0$. Since for a transverse wave we have $\mathbf{k} \cdot \mathbf{A}_0 = 0$, we are concerned with those directions of $\mathbf{p}$ for which $\mathbf{k} \cdot \mathbf{p} = 0$. Let us take the direction of $\mathbf{k}$ to be the x-axis of the electron co-ordinate system, and let us consider only that part of the interaction concerning the y-component of the electron momentum, p_y. We are then interested in the matrix element squared:

$$\left| \langle j | \frac{e}{m} p_y (1 + ik_x x) | i \rangle \right|^2. \tag{3.89}$$

In a formal sense there is a cross term to be dealt with in this square of a matrix element. But if $|i\rangle$ and $|j\rangle$ are states of well-defined parity the first-order (electric dipole) operator connects only states of opposite parity while the second-order operator connects states of the same parity. So the cross term vanishes. In any case, from an observational point of view, we are only interested in the small second-order multipole operator when the large first order effect vanishes in accordance with some selection rule. This is called a forbidden transition. Thus we consider

$$\left| \langle j | i \frac{e}{m} \frac{\omega}{c} p_y x | i \rangle \right|^2 \tag{3.90}$$

where we have put $k_x = \omega/c$.

The operator p_y commutes with x, so we may write $p_y x = x p_y$ and elaborate this operator as follows:

$$xp_y = \tfrac{1}{2}(xp_y - p_x y) + \tfrac{1}{2}(xp_y + p_x y)$$
$$= \tfrac{1}{2}\hbar l_z + \tfrac{1}{2}m(x\dot{y} + \dot{x}y) \tag{3.91}$$

where the first term is just the z-component of the orbital angular momentum. Its contribution to the matrix element (3.90) is

$$\frac{\omega^2}{c^2} \left| \langle j | \frac{e\hbar}{2m} l_z | i \rangle \right|^2. \tag{3.92}$$

But $(e\hbar/2m)l_z$ is just the z-component of the magnetic dipole moment of the atom (or rather, the orbital part of it; see chapter 4), so we speak of this contribution of the second order multipole moment as *magnetic dipole* radiation:

Magnetic dipole: $\qquad \dfrac{\omega^2}{c^2} \left| \langle j | \mu_{l_z} | i \rangle \right|^2 \tag{3.93}$

In the initial formulation of the radiation problem through eq. (3.25) we

49

omitted electron spin. Had we included it, we should have found that in lowest order it comes into the expression (3.93) in the form of the spin magnetic moment $\mu_{s_z} = g_s \mu_B s_z$ (see chapter 4) added to the orbital magnetic moment.

The second term of eq. (3.91) can be written

$$\tfrac{1}{2}m(x\dot{y} + \dot{x}y) = \tfrac{1}{2}m\frac{i}{\hbar}\,(x\mathcal{H}_0 y - xy\mathcal{H}_0 + \mathcal{H}_0 xy - x\mathcal{H}_0 y)$$

$$= \tfrac{1}{2}m\frac{i}{\hbar}\,(\mathcal{H}_0 xy - xy\mathcal{H}_0). \tag{3.94}$$

As in eq. (3.52) its contribution to the matrix element is

$$\frac{e^2}{m^2} \cdot \frac{\omega^2}{c^2} \cdot \frac{1}{4} \cdot \frac{m^2}{\hbar^2}\,(E_j - E_i)^2 |\langle j| \, xy \, |i\rangle|^2.$$

$$\text{Electric quadrupole:} \quad = \frac{1}{4}\frac{\omega^4}{c^2}\,|\langle j| \, exy \, |i\rangle|^2 \tag{3.95}$$

where $(E_j - E_i)/\hbar = \omega$. The operator in eq. (3.95) is a component of a second-rank tensor, the atomic electric quadrupole moment Q. Classically speaking, we see that, whereas the atomic electric dipole oscillates in phase with an incoming wave (no retardation), quadrupole radiation depends on the time of arrival of the electromagnetic disturbance at different parts of the electron charge distribution. Pure quadrupole radiation arises when two parts of the charge distribution are oscillating like electric dipoles out of phase so that the dipole contribution vanishes. Clearly for this kind of radiation to be important the wavelength of the incident wave must be comparable with the size of the charge distribution.

From eq. (3.90) it seems reasonable that the probability of magnetic dipole radiation is of the same order of magnitude as that of electric quadrupole radiation, for the two effects were formulated together. We can demonstrate this explicitly by comparing each with electric dipole radiation. From eq. (3.52) the square of an electric dipole matrix element is of the order of $\omega^2 d_e^2$ where $d_e < ea_0/Z$ is an atomic electric dipole moment. Similarly, we have for magnetic dipole, from eq. (3.93), $(\omega^2/c^2)\mu^2$; and for electric quadrupole, from eq. (3.95),

$$\omega^2\left(\frac{\omega^2}{c^2}Q^2\right).$$

Now

$$\frac{\mu/c}{d_e} \sim \frac{e\hbar}{mc} \cdot \frac{1}{ea_0/Z} = Z\alpha;$$

and

$$\frac{\omega Q/c}{d_e} \sim \frac{1}{4\pi\varepsilon_0}\frac{Z^2 e^2}{\hbar c a_0}\,e(a_0/Z)^2\,\frac{1}{ea_0/Z} = Z\alpha.$$

So

$$d_e^2 : \frac{\mu^2}{c^2} : \left(\frac{\omega^2 Q^2}{c^2}\right) = 1 : Z^2\alpha^2 : Z^2\alpha^2, \tag{3.96}$$

and both magnetic dipole and electric quadrupole effects are greatly reduced relative to electric dipole by the same factor $Z^2\alpha^2$. Magnetic dipole radiation, particularly, is important in induced transitions between states of the same configuration in the radiofrequency region where spontaneous transitions are negligible and electric dipole transitions are forbidden (the states involved are of the same parity). The whole field of radiofrequency spectroscopy in its various forms depends on this process. Spontaneous magnetic dipole and electric quadrupole transitions are mainly important in the X-ray region where the wavelength is of the same order as the size of an atom. These transitions also provide a weak mode of decay in the optical region for states from which decay by electric dipole radiation is forbidden. The lifetimes of such states against decay by second-order multipole radiation are correspondingly much longer than in the case of electric dipole radiation.

We shall not have occasion to go into the selection rules for these higher order radiations in detail, but they are mentioned again in chapter 7 in connection with general rules for many-electron atoms.

Problems

(Those problems marked with an asterisk are more advanced.)

3.1. Derive the formula $\omega^2/\pi^2 c^3$ for the number of modes per unit volume per unit frequency range for electromagnetic radiation confined to a cubical box.

3.2. Show that, with the gauge condition $\nabla \cdot \mathbf{A} = 0$ (eq. 3.32)), $\mathbf{p}$ commutes with $\mathbf{A}$ and hence $\mathbf{p} \cdot \mathbf{A} + \mathbf{A} \cdot \mathbf{p} = 2\mathbf{A} \cdot \mathbf{p}$.

3.3. Show that the term quadratic in $\mathbf{A}$ in the Hamiltonian of eq. (3.33) is small compared with the term linear in $\mathbf{A}$ for ordinary electric field strengths as follows: estimate the magnitude of $\mathbf{A}$ in terms of $\mathbf{E}$ from eq. (3.30) and use the approximation $p \sim \hbar/(a_0/Z)$ to show that $(e/m)\mathbf{A} \cdot \mathbf{p}$ is of the order of the interaction energy between an atomic dipole moment and the external field $\mathbf{E}$. Then show that the quadratic term $(e^2/2m)\mathbf{A}^2$ is less than the linear term by a factor $E/\{Z^3 e/4\pi\varepsilon_0 a_0^2\}$, and interpret this result.

3.4. Oscillator strengths occur in the elementary classical theory of the complex refractive index of a non-polar gas. To see the connection between this theory and that outlined in the text we consider the following problem. Let

$$\ddot{x} + \gamma\dot{x} + \omega_{ij}^2 x = -\frac{eE_0}{m} e^{i\omega t}$$

be the equation of motion of a one-dimensional damped oscillator, of charge $-e$, mass m, resonance frequency ω_{ij}, driven by an alternating electric field of amplitude E_0 and frequency ω. From the complex refractive index of a rarified gas of N_{ij} such oscillators per unit volume show that the absorption coefficient is, under sufficient approximation,

$$k_{ij} = \frac{\pi\varepsilon^2}{4\pi\varepsilon_0 mc} N_{ij} \frac{\gamma}{(\omega_{ij} - \omega)^2 + (\gamma/2)^2}$$

where k_{ij} is the absorption coefficient defined for intensities I, proportional to E_0^2, in the sense that an incident intensity I_0 is reduced to $I_s = I_0 e^{-k_{ij}s}$ after passing through a thickness s of the gas. If this model is to represent N one-electron *atoms* per unit volume each capable of absorption from the state i to many states j, we introduce the oscillator strength f_{ij} for absorption through

$$Nf_{ij} = N_{ij} \quad \text{or} \quad f_{ij} = \frac{N_{ij}}{N}$$

where

$$\sum_j f_{ij} = \sum_j \frac{N_{ij}}{N} = 1$$

to indicate that only a fraction f_{ij} of an oscillator's absorbing power is to be allocated to each channel $i \rightarrow j$ available to one atom. Summing over absorption channels we have

$$k_i = \frac{\pi\varepsilon^2 N}{4\pi\varepsilon_0 mc} \sum_j f_{ij} \frac{\gamma}{(\omega_{ij} - \omega)^2 + (\gamma/2)^2}$$

where f_{ij} appears as a simple weighting factor when the absorption coefficient is expressed in terms of N atoms rather than in terms of N_{ij} oscillators. This expression is a sum of Lorentz curves of absorption versus frequency. Show that the frequency width of an absorption line at half the peak absorption is γ.

Consider again just one absorption line $i \rightarrow j$. Let an incident plane wave of intensity $I_0(\omega) = \rho(\omega)c$ fall on the gas. The dimensions of the intensity are energy per unit frequency range crossing unit area per unit time; $\rho(\omega)$ is the energy density per unit frequency range introduced in eq. (3.2). $\rho(\omega)$ is an isotropic energy density, and in the above definition we are speaking of the *directed* intensity of a collimated beam, not of the intensity defined as energy flux *per unit solid angle*. When the sample of gas is 'optically thin', i.e., $k_{ij}s$ is small, show that the intensity $I_s(\omega)$ after passing through a thickness s of gas is

$$I_s(\omega) = I_0(\omega)(1 - k_{ij}s).$$

Problems

We define the 'equivalent width' w by integrating over all frequencies of the incident wave:

$$w = \int_0^\infty \frac{I_0(\omega) - I_s(\omega)}{I_0(\omega)}\, d\omega = s \int_0^\infty k_{ij}\, d\omega$$

where $\int_0^\infty k_{ij}\, d\omega$ is called the total absorption for the line $i \to j$. Interpret the meaning of the term 'equivalent width'. Show that

$$\int_0^\infty k_{ij}\, d\omega = \frac{\pi e^2}{2\varepsilon_0 mc}\, N f_{ij}.$$

Notice the important result that the area under the absorption curve is independent of γ, and in general of the mechanism of broadening of the line. The broadening mechanism determines the shape of the line, not the area integrated over frequency.

Show that if $I_0(\omega)$ varies only slowly with frequency over the region of the absorption line the rate of loss of energy from the incident beam per unit volume of gas by absorption is just $N\rho(\omega)B_{ij}\hbar\omega_{ij}$—as in eq. (3.2)—provided that eq. (3.76) holds for the relation between f_{ij} and B_{ij}.

For further reading see, for example, A. C. G. Mitchell and M. W. Zeemansky, *Resonance Radiation and Excited Atoms*; H. G. Kuhn, *Atomic spectra*, Second Edition, sections II D, VII B, F.

*3.5. Prove the f-sum rule for absorption as follows: consider the matrix element, eq. (3.52),

$$\langle j| \frac{e}{m} p_x |i\rangle = \frac{i}{\hbar}(E_j - E_i)\langle j| ex |i\rangle.$$

and use it to evaluate, by matrix multiplication, the diagonal matrix element of the commutator $p_x x - x p_x$:

$$\langle i| p_x x - x p_x |i\rangle.$$

The commutation relation $p_x x - x p_x = -i\hbar$ gives the result in the form of eq. (3.76)

$$\sum_j \frac{2}{3} \cdot \frac{m}{e^2\hbar^2}(E_j - E_i)\,|\langle i| er |j\rangle|^2 = 1$$

which is the sum rule.

*3.6. Use the Schrödinger hydrogenic wave functions for $n = 1$, $l = 0$, $m_l = 0$ and for $n' = 2$, $l' = 1$, $m_l' = 1, 0, -1$ to show that

$$|\langle 210| z |100\rangle|^2 = \frac{1}{2}|\langle 21 \pm 1| x \pm iy |100\rangle|^2 = \frac{2^{15}}{3^{10}}\, a_0^2.$$

Evaluate the Einstein A coefficient for the $1s$—$2p$ transition (Lyman α line) and hence show that the lifetime of the $2p$ level against spontaneous decay by electric dipole radiation is $\tau = 1\cdot6 \times 10^{-9}$ s. (Be careful about the definition of A in terms of matrix elements when the upper level is degenerate.)

3.7. Given that the lifetime of the $2p$ level of hydrogen against decay by electric dipole radiation is $\tau = 1.6 \times 10^{-9}$s, estimate the lifetime of the $2p$ level of (a) hydrogen-like Ti^{21+}; (b) muonic titanium (see problem **2.9**).

*3.8. Consider an atom subject to an electromagnetic field which is capable of inducing transitions between two of its states only—no other states are involved. Such a system is often called a *two-level atom*. Let these states p and q satisfy the Schrödinger equation (3.36)

$$\mathcal{H}_0 \,|p\rangle = E_p|p\rangle; \qquad \mathcal{H}_0|q\rangle = E_q|q\rangle;$$

so that the resonance frequency for transitions between them is $\omega_0 = (E_q - E_p)/\hbar$. Assume that the perturbation $\mathcal{H}'$, representing the interaction with a monochromatic field of frequency ω, has matrix elements

$$\langle p| \,\mathcal{H}'\,|q\rangle = \hbar b\, e^{i\omega t}; \qquad \langle q| \,\mathcal{H}'\,|p\rangle = \hbar b\, e^{-i\omega t};$$

$$\langle p| \,\mathcal{H}'\,|p\rangle = \langle q| \,\mathcal{H}'\,|q\rangle = 0.$$

This assumption reflects the rotating wave approximation: if the applied field has a time-dependence $\cos \omega t$, only one rotating component is effective in inducing both absorption and induced emission.

Spontaneous emission may be neglected if $b \gg \gamma$ where γ is the frequency width of the transition due to spontaneous emission. This condition may come about either because the radiation field is very intense, as in some laser experiments, or because γ^{-1} is long compared with the duration of the experiment.

(a) Estimate, with the help of eq. (3.57), the values of γ^{-1} for a spontaneous magnetic dipole transition between two states whose frequency separation is 1,000 MHz.

(b) Use eq. (3.38) to set up the following equations of motion for the amplitudes $c_p(t)$ and $c_q(t)$ (defined by eq. (3.35)):

$$i\dot{c}_p = c_q b\, e^{-i(\omega_0 - \omega)t},$$

$$i\dot{c}_q = c_p b\, e^{i(\omega_0 - \omega)t}.$$

(c) Assuming that $c_p(0) = 1$, $c_q(0) = 0$, solve these equations of motion exactly to show that the probability of transition to state q after a time t is

$$P_{pq} = |c_q(t)|^2 = \frac{(2b)^2}{\Omega^2} \sin^2 \frac{1}{2} \Omega t,$$

where $\Omega^2 = (\omega_0 - \omega)^2 + (2b)^2$. This is often called the Rabi formula.

(d) At resonance P_{pq} varies as $\sin^2 bt$, hence b is called the Rabi flopping frequency. Show that, for the case of electric dipole radiation, $b = d_e E_0/2\hbar$ where d_e is the atomic electric dipole matrix element and the applied field is $\mathbf{E}(t) = \hat{\mathbf{e}} E_0 \cos \omega t$. Show that an analogous expression is obtained for magnetic dipole radiation.

(e) In an atomic-beam magnetic-resonance experiment (see N. F. Ramsey, *Molecular Beams*, chapter V, O.U.P. 1956) an oscillating magnetic field is applied for a fixed time τ (as experienced by atoms passing through the interaction region with a given speed). Show that there is an optimum value of b which maximises P_{pq} and find this maximum value of P_{pq}.

(f) Show that the factor $(2b)^2/\Omega^2$ has a full frequency width at half-maximum intensity of $4b$. This effect is called power broadening. Discuss the difference between the exact treatment set out in this problem and the low-signal first-order solution of eq. (3.43).

(g) Note that the expansion coefficients defined through eq. (3.35) could have been defined in a different way:

$$\Psi = \sum_n C_n(t)\psi_n$$

in which the wave function ψ_n is independent of time and the factor $e^{-iE_n t/\hbar}$ has been absorbed in $C_n(t)$. Show that the equations of motion replacing those of part (b) are now

$$i\dot{C}_p = C_p E_p/\hbar + C_q b\, e^{i\omega t}$$
$$i\dot{C}_q = C_q E_q/\hbar + C_p b\, e^{-i\omega t}.$$

The final answer for $|C_q(t)|^2$ is the same as for $|c_q(t)|^2$ because C_q and c_q differ only by a phase factor which does not matter in the result.

*3.9. Consider again the two-level atom of problem **3.8** subject to a perturbation of Rabi flopping frequency b. In this problem we want to include the possibility of decay by spontaneous emission at a rate γ from the upper state to the lower state. Let q be the upper and p the lower state.

(a) Confirm that a plausible way of introducing this phenomenon is to replace the equations of motion (part (b) of problem **3.8**) by

$$i\dot{c} = i(\gamma/2)c_q + c_q b\, e^{-i(\omega_0 - \omega)t}$$
$$i\dot{c}_q = -i(\gamma/2)c_q + c_p b\, e^{i(\omega_0 - \omega)t},$$

provided that $|c_p(t)|^2$ departs very little from its initial value $|c_p(0)|^2 = 1$ (otherwise the normalisation condition $|c_p(t)|^2 + |c_q(t)|^2 = 1$ cannot be satisfied). We therefore treat these equations only by perturbation theory.

(b) Show that in first-order perturbation theory the probability of excitation of the state q after a time t is

$$|c_q(t)|^2 = \frac{b^2}{(\omega_0 - \omega)^2 + (\gamma/2)^2} \{1 - 2\,e^{-(\gamma/2)t}\cos(\omega_0 - \omega)t + e^{-\gamma t}\}.$$

This formula is valid for $b \ll \gamma$.

As in the classical treatment (problem 3.4), the first factor in this formula is a Lorentzian function of frequency with width γ.

4. The hydrogen atom: fine structure

The Schrödinger theory with which we have treated hydrogen so far is inadequate to explain certain details in the spectrum of hydrogen and of other simple atoms. We can quote three pieces of experimental spectroscopic evidence which historically gave rise to difficulty. These difficulties were resolved with the introduction of the concept of *electron spin*.

4.1. Electron spin

(a) The Stern–Gerlach experiment was first done by passing a highly collimated beam of silver atoms in vacuo through a region in which there is a strong gradient of magnetic field, $\partial B/\partial z$, in the z-direction transverse to the beam. (See problem 4.1.) If the atom has a magnetic moment μ then there is a force on the atom of magnitude $\mu_z(\partial B/\partial z)$, and a transverse deflection of the beam may be observed in the plane of a detector. Classically, without space quantization, a continuous distribution of deflections would have been expected, but this is not what was observed.

A single-electron atom has a magnetic moment associated with the orbital motion of the electron. Semi-classically the following expression is derived:

$$\boldsymbol{\mu}_l = \oint i \tfrac{1}{2}\mathbf{r} \times \mathbf{ds} \tag{4.1}$$

where i is the electron current in orbit, ds is an element of orbital path and $\tfrac{1}{2}\mathbf{r} \times \mathbf{ds}$ is an element of area. This becomes

$$\boldsymbol{\mu}_l = \oint \tfrac{1}{2}\mathbf{r} \times \frac{\mathbf{ds}}{\mathrm{d}t}\, \mathrm{d}q$$

$$= \frac{1}{2m_0} \oint \mathbf{r} \times \mathbf{p}\, \mathrm{d}q, \tag{4.2}$$

where $\mathrm{d}q$ is an element of electron charge. But $\mathbf{r} \times \mathbf{p}$ is the orbital angular momentum which is a constant of the motion, $\hbar\mathbf{l}$, so it may be taken outside the integral:

$$\boldsymbol{\mu}_l = \frac{1}{2m_0}\,\hbar\mathbf{l}(-e) \tag{4.3}$$

where $-e$ is the total electron charge, or

$$\boldsymbol{\mu}_l = -\frac{e\hbar}{2m_0}\mathbf{l} \equiv -g_l\mu_B\mathbf{l}, \tag{4.4}$$

where $\mu_B = e\hbar/2m_0 = 9.2732 \times 10^{-24}\ \mathrm{JT}^{-1}$ is the Bohr magneton. g_l is simply a generalized constant of proportionality between $\boldsymbol{\mu}_l$ and $-\mu_B\mathbf{l}$: for orbital motion $g_l = 1$.

Now $\mu_{l_z} = -g_l\mu_B m_l$ where m_l can take $2l + 1$ values. Since l is integral, $2l + 1$ is odd. If the magnetic moment of an atom were associated only with orbital motion there would be an odd number of discrete traces in the deflection pattern of a Stern–Gerlach experiment. Stern and Gerlach confirmed the discrete nature of space quantization in a very direct manner, but for silver they found *two* traces symmetrically disposed on either side of the beam axis, with *no* undeflected trace.

(b) The second piece of evidence also concerns space quantization, but this time in a spectroscopic transition: when a magnetic field is applied to a light source the resulting Zeeman effect in the spectral lines is in general the so-called 'anomalous' effect which, in contrast to the 'normal' effect, cannot be understood in terms of orbital motion alone.

(c) Thirdly, in the alkali metals many of the lines show a well-resolved *doublet* fine structure. For example, the well-known yellow D-lines of sodium are a doublet with a separation of about $17\ \mathrm{cm}^{-1}$.

For all these reasons what appears to be needed is another quantum number for which the number of projections in space quantization is even. Uhlenbeck and Goudsmit postulated spin angular momentum empirically. (Historically their proposal came towards the end of a very fruitful period, up to 1924, during which the Old Quantum Theory was being refined, and just before the introduction of wave-mechanics and matrix-mechanics in 1925 and 1926.)

In quantum mechanics we speak of spin angular momentum, $\hbar\mathbf{s}$. $\mathbf{s}$ operates on a spin eigenfunction χ in a special space quite separate from co-ordinate space such that:

$$\mathbf{s}^2\chi = s\,(s + 1)\chi \tag{4.5}$$

and

$$s_z\chi = m_s\chi \tag{4.6}$$

where $s = \tfrac{1}{2}$ and $m_s = \pm\tfrac{1}{2}$ only. Thus there are only two, i.e., $(2s + 1)$, projections of $\mathbf{s}$ and associated with them only two normalized and orthogonal eigenfunctions $\chi(+)$ and $\chi(-)$ defined by

$$s_z\chi(+) = \tfrac{1}{2}\chi(+) \tag{4.7}$$

$$s_z\chi(-) = -\tfrac{1}{2}\chi(-); \tag{4.8}$$

$$\chi^*(+)\chi(+) = \chi^*(-)\chi(-) = 1, \tag{4.9}$$

$$\chi^*(+)\chi(-) = \chi^*(-)\chi(+) = 0. \tag{4.10}$$

We introduce these spin functions here merely as functions on which the spin angular momentum operates. Formally they are playing the same role as the spherical harmonics played in eq. (2.68) and (2.69) for orbital angular momentum, but because of their double-valued nature they have rather special properties (see also problem 4.2). We shall need them later to give a quantum-mechanical description of a system which has spin.

The spin **s** also obeys commutation rules of the form

$$[s_x, s_y] = is_z \quad \text{(and cyclically)}, \tag{4.11}$$

rules which can be regarded as being part of the definition of an angular momentum operator (see appendix C). Notice that there is no integration in the normalization condition (4.9), and there is no approach to the classical limit in the sense that $s \to \infty$ because s is confined to the value $\tfrac{1}{2}$ only. So spin is entirely a quantum-mechanical concept. It is not included naturally in the Schrödinger treatment but is provided for by Dirac's relativistic electron theory.

Associated with the spin angular momentum is a spin magnetic moment, which by analogy with eq. (4.4) is

$$\boldsymbol{\mu}_s = -g_s\mu_B\mathbf{s}. \tag{4.12}$$

According to Dirac the spin g-factor is

$$g_s = 2, \tag{4.13}$$

(not 1 as for orbital motion) which is the value required for a quantitative fit to the spectroscopic data.

If we stick to the Schrödinger treatment but modify it by the introduction of spin as an additional concept with *no interaction* between spin and orbit, we can form in this approximation a new wave function, including spin, which is separable in space and spin variables:

$$\psi_{n, l, m_l, m_s} = \psi_{n, l, m_l}\chi(m_s). \tag{4.14}$$

We now have four quantum numbers to describe a one-electon atom: n, l, m_l and m_s.

4.2. The interaction terms

The fine structure of the energy levels of hydrogen is due to relativistic effects which are properly treated by the Dirac equation. But since the effects are very small one can approximate the Dirac equation to the non-relativistic case, using Schrödinger functions (4.14) as the zeroth-order functions in perturbation theory, and keeping only terms up to order v^2/c^2 in an expansion of the Dirac Hamiltonian in powers of v/c. In addition to the kinetic energy and Coulomb potential energy terms in the Schrödinger equation there are three small terms which can be treated by first-order perturbation theory. Since an investigation of the Dirac Hamiltonian is beyond the scope of this book we shall simply quote these terms.

(a) The first term is

$$-\frac{1}{2m_0c^2}(E_n - V(r))^2.$$

This term does not contain spin, and can be arrived at by a modification of the Schrödinger Hamiltonian $\mathscr{H}_0$ in a way which describes the relativistic variation of mass.

The relativistic form of the total energy for a central field is

$$\mathscr{H} = (p^2c^2 + m_0^2c^4)^{1/2} + V(r) \qquad (4.15)$$

from which the kinetic energy T is

$$T = (p^2c^2 + m_0^2c^4)^{1/2} - m_0c^2$$

$$\approx \frac{p^2}{2m_0} - \frac{1}{8}\cdot\frac{p^4}{m_0^3c^2} + \cdots$$

$$= T_0 - \frac{1}{8}\cdot\frac{p^4}{m_0^3c^2}\cdots \qquad (4.16)$$

If we write

$$-\frac{1}{8}\cdot\frac{p^4}{m_0^3c^2} = -\frac{1}{2m_0c^2}\left(\frac{p^2}{2m_0}\right)^2$$

$$= -\frac{1}{2m_0c^2}T_0^2$$

$$= -\frac{1}{2m_0c^2}(E_n - V(r))^2, \qquad (4.17)$$

we obtain a modification to the non-relativistic kinetic energy T_0 which is the same as the Dirac result to second order in v/c.

Using the Schrödinger functions to evaluate the energy shift in first-order perturbation theory, we have

$$\Delta E_n' = -\frac{1}{2m_0c^2}\langle(E_n - V(r))^2\rangle$$

$$= -\frac{1}{2m_0c^2}\left\{E_n^2 - 2E_n\left\langle-\frac{Ze^2}{4\pi\varepsilon_0 r}\right\rangle + \left\langle\frac{Z^2e^4}{(4\pi\varepsilon_0)^2r^2}\right\rangle\right\}. \qquad (4.18)$$

4.2. The interaction terms

Notice that the fact that there is degeneracy with respect to l in the zeroth-order energy levels E_n is of no consequence here because the perturbation is itself diagonal in l. We can evaluate $\Delta E_n'$ using

$$\langle r^{-1} \rangle = \frac{1}{n^2} \cdot \frac{Z}{a_0}, \qquad \langle r^{-2} \rangle = \frac{1}{(l + \frac{1}{2})n^3} \left(\frac{Z}{a_0} \right)^2$$

from table 2.3, and

$$E_n = -\frac{1}{4\pi\varepsilon_0} \frac{Z^2 e^2}{2n^2 a_0}$$

from eq. (2.57):

$$\Delta E_n' = -\frac{\alpha^2 Z^2}{n^2} E_n \left(\frac{3}{4} - \frac{n}{l + \frac{1}{2}} \right). \qquad (4.19)$$

This term now depends on l as well as on n. It is smaller than E_n by a factor $\alpha^2 Z^2 \sim v^2/c^2$.

  (b) The next term is

$$\frac{1}{2m_0 c^2} \left(\frac{e\hbar}{2m_0} \right) 2\mathbf{s} \cdot (\mathbf{E} \times \mathbf{p}).$$

This term is spin-dependent. $\mathbf{E}$ is the electric field through which the electron is moving: $\mathbf{E} = -\nabla\phi$ where $\phi = Ze/4\pi\varepsilon_0 r$ or $\mathbf{E} = (1/e)\nabla V$ where $V = -Ze^2/4\pi\varepsilon_0 r$. Since $V = V(r)$ we can write an expression which will serve for the more general case of a central field:

$$\mathbf{E} = \frac{1}{e} \frac{\mathbf{r}}{r} \frac{dV}{dr}. \qquad (4.20)$$

Then

$$\frac{1}{2m_0 c^2} \left(\frac{e\hbar}{2m_0} \right) 2\mathbf{s} \cdot (\mathbf{E} \times \mathbf{p}) = \frac{1}{2m_0 c^2} \frac{\hbar^2}{2m_0} 2\mathbf{s} \cdot \left(\frac{\mathbf{r} \times \mathbf{p}}{\hbar} \right) \frac{1}{r} \frac{dV}{dr}$$

$$= \frac{\hbar^2}{2m_0^2 c^2} \frac{1}{r} \frac{dV}{dr} \mathbf{s} \cdot \mathbf{l}. \qquad (4.21)$$

From its form this term is called the spin–orbit interaction, and it arises from the relativistic motion (orbital) of an electron with spin through an electric field whose source is the nuclear charge.

  We can estimate the spin–orbit interaction with the electron spin, and its magnetic moment, introduced empirically into the non-relativistic theory. We add a magnetic interaction term as a perturbation:

$$\mathcal{H}' = -\boldsymbol{\mu}_s \cdot \mathbf{B}, \qquad (4.22)$$

where $\mathbf{B}$ is the magnetic field which the electron experiences as a result of moving through the electric field **E**.

$$\mathbf{B} = \frac{1}{c^2}\,\mathbf{E} \times \mathbf{v} = \frac{1}{m_0 c^2}\,(\mathbf{E} \times \mathbf{p}). \tag{4.23}$$

Then, since $\boldsymbol{\mu}_s = -2\mu_B \mathbf{s}$,

$$-\boldsymbol{\mu}_s \cdot \mathbf{B} = \frac{2\mu_B}{m_0 c^2}\,\mathbf{s} \cdot (\mathbf{E} \times \mathbf{p})$$

$$= 2\,\frac{1}{2m_0 c^2}\left(\frac{e\hbar}{2m_0}\right) 2\mathbf{s} \cdot (\mathbf{E} \times \mathbf{p}). \tag{4.24}$$

This is just twice as large as the Dirac term. The calculation is incomplete, however, because it has been conducted in the rest frame of the moving electron rather than in that of the central charge. The so-called Thomas precession is a relativistic kinematic effect which modifies eq. (4.24) by the addition of a term which happens to be $-\frac{1}{2}$ as large as the semi-classical spin–orbit term, and thus brings agreement with the Dirac expression.

The first-order energy shift arising from the spin–orbit interaction is

$$\Delta E'' = \frac{\hbar^2}{2m_0^2 c^2}\left\langle \frac{1}{r}\frac{dV}{dr}\,\mathbf{s} \cdot \mathbf{l} \right\rangle. \tag{4.25}$$

With $V = -Ze^2/4\pi\varepsilon_0 r$,

$$\frac{1}{r}\frac{dV}{dr} = \frac{Ze^2}{4\pi\varepsilon_0 r^3} \tag{4.26}$$

and from table 2.3

$$\langle r^{-3} \rangle = \frac{Z^3}{a_0^3 n^3 l(l + \frac{1}{2})(l + 1)},$$

so

$$\Delta E'' = -\frac{\alpha^2 Z^2}{n^2}\,E_n\,\frac{n}{l(l + \frac{1}{2})(l + 1)}\,\langle \mathbf{s} \cdot \mathbf{l} \rangle,\, l \neq 0. \tag{4.27}$$

To maintain continuity in the exposition we shall anticipate the result, eq. (4.58), for $\langle \mathbf{s} \cdot \mathbf{l} \rangle$:

$$\langle \mathbf{s} \cdot \mathbf{l} \rangle = \tfrac{1}{2}\{j(j + 1) - l(l + 1) - s(s + 1)\} \tag{4.28}$$

where $s = \frac{1}{2}$ and j is a new quantum number describing the total angular momentum. The allowed values of j in this case are $l \pm \frac{1}{2}$. With eq. (4.28) for $\langle \mathbf{s} \cdot \mathbf{l} \rangle$ we can combine eqs. (4.19) and (4.27) to give

$$\Delta E' + \Delta E'' = -\frac{\alpha^2 Z^2}{n^2}\,E_n\left\{\frac{3}{4} - \frac{n}{j + \frac{1}{2}}\right\},\quad j = l \pm \tfrac{1}{2},\, l \neq 0. \tag{4.29}$$

4.2. The interaction terms

(c) The third term is

$$-\frac{\hbar^2}{4m_0^2 c^2} e\mathbf{E} \cdot \nabla,$$

and it applies only to the special case of $l = 0$. This case always gives difficulty because the wave function does not vanish at the origin. In many problems the electronic behaviour is then dominated by the region of very small r for which the non-relativistic approximation $Ze^2/4\pi\varepsilon_0 r \ll m_0 c^2$ is very seriously violated. This third term is the so-called Darwin term which has no classical analogue. We now quote the energy shift derived from it (see problem 4.3):

$$\Delta E''' = \frac{\pi\hbar^2}{2m_0^2 c^2} \frac{Ze^2}{4\pi\varepsilon_0} |\psi(0)|^2, \quad l = 0. \tag{4.30}$$

The appearance of $|\psi(0)|^2$ in eq. (4.30) underlines the fact that this formula is only applicable to the case $l = 0$. From eqs. (4.30) and (2.56) we obtain

$$\Delta E''' = -\frac{\alpha^2 Z^2}{n^2} E_n n. \tag{4.31}$$

Because the spin–orbit interaction does not apply when $l = 0$ we must put $\Delta E'' = 0$ for this case, and the total shift is $\Delta E' + \Delta E'''$; but for $l \neq 0$, $\Delta E''' = 0$ and the total shift is $\Delta E' + \Delta E''$. We can write a single formula, valid for all l, for the total shift ΔE:

$$\Delta E = \Delta E' + \Delta E'' + \Delta E''' = -\frac{\alpha^2 Z^2}{n^2} E_n \left(\frac{3}{4} - \frac{n}{j + \frac{1}{2}}\right), \quad j = l \pm \frac{1}{2}. \tag{4.32}$$

The right-hand side of this equation turns out to be the same as that of eq. (4.29), but now the total expression applies to all l, including $l = 0$. Putting in the expression (2.57) for E_n we obtain the full dependence of the fine-structure energy shift on Z and n:

$$\Delta E_{n,j} = \alpha^2 \frac{m_0}{2\hbar^2} \frac{e^4}{(4\pi\varepsilon_0)^2} \frac{Z^4}{n^4} \left(\frac{3}{4} - \frac{n}{j + \frac{1}{2}}\right). \tag{4.33}$$

α is called the fine-structure constant because it fixes the order of magnitude of ΔE relative to E_n ($\alpha \sim \frac{1}{137}$ and $\alpha^2 \sim 5 \times 10^{-5}$).

The Dirac equation for this problem can be solved exactly, and eq. (4.33) is the approximate form of the exact solution up to terms in v^2/c^2.

In its fine structure, as in its gross structure, hydrogen is a special case, not at all typical of alkali atoms with one valence electron outside a core of spherically symmetrical shells. The most important feature in hydrogen is that the fine structure is so small that all three of the relativistic effects which we have quoted in this section have the same order of magnitude, so they must all be treated together. For many-electron atoms we shall find that the spin–orbit effect (the second term here) emerges as the largest term and the other relativistic corrections can be ignored. The final result, eq. (4.33), *does not depend on l*, even though the three separate contributions do. (In the full relativistic treatment **l** is not a constant of the motion because we can say that there is a torque on it due to the spin, but in the zeroth-order wave functions which we have used for first-order perturbation theory we treat l as a good quantum number.) Thus the degeneracy with respect to l in the gross structure is not lifted in the fine structure.

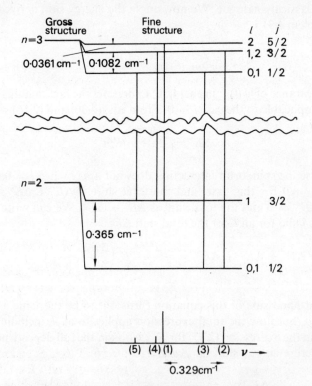

Fig. 4.1. Fine structure of the $n = 2$ and $n = 3$ levels of hydrogen according to the Dirac theory, with the spectrum of allowed transitions.

There is a splitting into levels labelled by different j, but levels of the same j and different l for a given n coincide: $E_n(j = l + \frac{1}{2}) = E_n(j = l' - \frac{1}{2})$ where $l' = l + 1$. Figure 4.1 shows the fine structure of the $n = 2$ and $n = 3$ terms of hydrogen with the transitions which give rise to the fine structure of the Balmer line H_α according to the Dirac theory. The electric dipole selection rules for j, which we shall discuss in more detail later, are

$$\Delta j = 0, \pm 1. \qquad (4.34)$$

4.3. The vector model

The structure of the $n = 3$ term is considerably smaller than that of $n = 2$ because $\Delta E_{n,j} - \Delta E_{n,j'} \propto n^{-3}$. We also see from eq. (4.33) that $\Delta E_{n,j} \propto Z^4$, so the fine structure for a given n increases rapidly along the iso-electronic sequence H, He$^+$, Li^{++}, $\cdots$

The main feature of Fig. 4.1 is the appearance of two strong components, (1) and (2), separated approximately by the fine structure splitting of the $n = 2$ term, ~ 0.365 cm^{-1}. This structure is so small that resolution of the other components by optical methods is a difficult matter experimentally, especially so because the Doppler width in hydrogen is relatively large. We shall discuss the fine structure of H$_\alpha$ further in section 4.4.

4.3. The vector model

In this section, which will be useful for later work, we fill in the steps which led to eq. (4.28). In that equation $\xi \mathbf{s} \cdot \mathbf{l}$ was treated as a small perturbation, where

$$\xi = \frac{\hbar^2}{2m_0^2 c^2} \left\langle \frac{1}{r} \frac{dV}{dr} \right\rangle.$$

The total Hamiltonian is

$$\mathscr{H} = \mathscr{H}_0 + \xi \mathbf{s} \cdot \mathbf{l}, \tag{4.35}$$

where

$$\mathscr{H}_0 = -\frac{\hbar^2}{2m} \nabla^2 - \frac{Ze^2}{4\pi\varepsilon_0 r}. \tag{4.36}$$

The eigenfunctions of the zeroth-order Hamiltonian $\mathscr{H}_0$ are product functions in space and spin because in zeroth order there is no interaction between spin and orbit:

$$u_{n,l,m_l,m_s} = R_{nl} Y_l^{m_l} \chi(m_s). \tag{4.37}$$

For a given n these are all degenerate with respect to l, m_l and m_s, so we have to use degenerate perturbation theory (see appendix B). The necessary step, namely that of finding a new representation in which the perturbation $\xi \mathbf{s} \cdot \mathbf{l}$ is diagonal, is equivalent to finding the constants of the motion for the dynamical problem in which eq. (4.35) is the Hamiltonian. u_{n,l,m_l,m_s} is an eigenfunction of $\mathbf{l}^2$, $\mathbf{s}^2$, l_z, and s_z, that is, these operators commute with $\mathscr{H}_0$. With the perturbation added, $\mathbf{l}^2$ and $\mathbf{s}^2$ still commute with $\mathscr{H}$, but l_z and s_z do not. (We are only concerned with an angular problem here. In first-order perturbation theory the radial function R_{nl} can be used to find the expectation value of the radial part of the operator $\langle 1/r \cdot dV/dr \rangle$. This is already implicit in ξ.)

To investigate the constants of the motion we discuss the *vector model* which is a pictorial classical description of the behaviour of angular

65

momentum vectors. The classical equation of motion for the orbital angular momentum vector $\hbar\mathbf{l}$ interacting with the spin $\hbar\mathbf{s}$ is

$$\frac{\mathrm{d}}{\mathrm{d}t}(\hbar\mathbf{l}) = \zeta\mathbf{s} \times \mathbf{l} \tag{4.38}$$

for there is a torque $\zeta\mathbf{s} \times \mathbf{l}$ acting on $\hbar\mathbf{l}$. On the other hand the quantum-mechanical equation of motion, eq. (2.74), is

$$\frac{\mathrm{d}}{\mathrm{d}t}\langle\hbar\mathbf{l}\rangle = \frac{1}{\mathrm{i}\hbar}\langle[\hbar\mathbf{l}, \mathscr{H}]\rangle \tag{4.39}$$

which can be extended to be an equation in the operators themselves, rather than in their expectation values:

$$\frac{\mathrm{d}}{\mathrm{d}t}(\hbar\mathbf{l}) = \frac{1}{\mathrm{i}\hbar}[\hbar\mathbf{l}, \mathscr{H}]. \tag{4.40}$$

Let us work this out for the x-component of $\mathbf{l}$. All the components of $\mathbf{l}$ commute with $\mathscr{H}_0$, so we only need

$$\frac{\mathrm{d}}{\mathrm{d}t}(\hbar l_x) = -\mathrm{i}[l_x, \zeta\mathbf{s} \cdot \mathbf{l}]. \tag{4.41}$$

Remembering the commutation rules for $\mathbf{l}$,

$$[l_x, l_y] = \mathrm{i}l_z, \tag{4.42}$$

and the fact that l_x commutes with all components of $\mathbf{s}$ because these operators act in different spaces, we have

$$\begin{aligned}
\frac{\mathrm{d}}{\mathrm{d}t}(\hbar l_x) &= -\mathrm{i}\xi[l_x, (s_y l_y + s_z l_z)] \\
&= -\mathrm{i}\xi(\mathrm{i}s_y l_z - \mathrm{i}s_z l_y) \\
&= \xi(\mathbf{s} \times \mathbf{l})_x
\end{aligned} \tag{4.43}$$

and similarly for l_y and l_z, so

$$\frac{\mathrm{d}}{\mathrm{d}t}(\hbar\mathbf{l}) = \zeta\mathbf{s} \times \mathbf{l}. \tag{4.44}$$

But this is just the same as the classical equation (4.38) if the classical vectors replace the quantum-mechanical operators, so the vector model is a good description in this context.

Because $\mathbf{s}$ obeys the commutation relations

$$[s_x, s_y] = \mathrm{i}s_z \quad \text{(and cyclically)} \tag{4.45}$$

it has an equation of motion

$$\frac{\mathrm{d}}{\mathrm{d}t}(\hbar\mathbf{s}) = \zeta\mathbf{l} \times \mathbf{s}. \tag{4.46}$$

4.3. The vector model

If we add eqs. (4.44) and (4.46) we form a new angular momentum operator, the total angular momentum

$$\hbar \mathbf{j} = \hbar(\mathbf{l} + \mathbf{s}) \tag{4.47}$$

whose equation of motion is

$$\frac{\mathrm{d}}{\mathrm{d}t}(\hbar \mathbf{j}) = \xi \mathbf{s} \times \mathbf{l} + \xi \mathbf{l} \times \mathbf{s} = 0. \tag{4.48}$$

In other words, $\mathbf{j}$ is a constant of the motion for the problem and it commutes with the Hamiltonian $\mathscr{H}_0 + \xi \mathbf{s} \cdot \mathbf{l}$. This constant of the motion is just what we have set out to find.

It can be verified that $\mathbf{j}$, introduced here through eq. (4.47), obeys the commutation rules

$$[j_x, j_y] = \mathrm{i}j_z \quad \text{(and cyclically)} \tag{4.49}$$

and so, like $\mathbf{l}$ and $\mathbf{s}$, it satisfies the rule for a quantum-mechanical angular momentum operator.

The mechanics of the vector model are as follows: since $\mathbf{l} \times \mathbf{l} = 0$ we can rewrite eq. (4.38) as

$$\frac{\mathrm{d}}{\mathrm{d}t}(\hbar \mathbf{l}) = \xi(\mathbf{s} + \mathbf{l}) \times \mathbf{l} = \xi \mathbf{j} \times \mathbf{l}. \tag{4.50}$$

Similarly

$$\frac{\mathrm{d}}{\mathrm{d}t}(\hbar \mathbf{s}) = \xi \mathbf{j} \times \mathbf{s}. \tag{4.51}$$

These two equations represent a classical *precession* of $\mathbf{l}$ and $\mathbf{s}$ about $\mathbf{j}$ with a precession frequency $\omega = |\xi \mathbf{j}|/\hbar$, i.e.,

$$\frac{\mathrm{d}\mathbf{l}}{\mathrm{d}t} = \boldsymbol{\omega} \times \mathbf{l}. \tag{4.52}$$

Fig. 4.2. The vector model representing $\mathbf{l} + \mathbf{s} = \mathbf{j}$, with precession.

The vector addition $\mathbf{l} + \mathbf{s} = \mathbf{j}$ and the precession are represented in Fig. 4.2. The larger the magnitude of ξ, which determines the energy of interaction in the perturbation term $\xi \mathbf{s} \cdot \mathbf{l}$, the faster the rate of precession in the model because $\omega \propto \xi$. Time averages in the vector model become equivalent to expectation values in quantum mechanics. Thus the lengths $|\mathbf{l}|$, $|\mathbf{s}|$, and $|\mathbf{j}|$ are constant in the model ($\mathbf{l}^2$, $\mathbf{s}^2$, and $\mathbf{j}^2$ commute with the Hamiltonian) but the projections l_z and s_z on a z-axis fixed in space fluctuate during the precessional motion (m_s and m_l are not good quantum numbers and u_{n, l, m_l, m_s} is not a good representation). The larger the perturbation $\xi \mathbf{s} \cdot \mathbf{l}$ the more rapid the fluctuation and the worse the m_l, m_s representation. By contrast j and its projection m_j are good quantum numbers—there is no external torque on the system as a whole arising from the interaction $\xi \mathbf{s} \cdot \mathbf{l}$ which is internal to the atom.

We now return to the problem of finding the first-order energy shift arising from the perturbation $\xi \mathbf{s} \cdot \mathbf{l}$ in degenerate perturbation theory. For the zeroth-order wave functions we have to use a representation in which $\mathbf{s} \cdot \mathbf{l}$ is diagonal. The functions u_{n, l, m_l, m_s} will not do because $\mathbf{s} \cdot \mathbf{l}$ does not commute with l_z or s_z; but we can choose satisfactory zeroth-order functions v_{n, l, j, m_j} by forming certain linear combinations of the basis set u_{n, l, m_l, m_s}. The labels of v_{n, l, j, m_j} indicate that these functions are simultaneous eigenfunctions of $\mathbf{l}^2$ (and $\mathbf{s}^2$), $\mathbf{j}^2$ and j_z. They are satisfactory because $\mathbf{s} \cdot \mathbf{l}$ commutes with $\mathbf{l}^2$, $\mathbf{s}^2$, $\mathbf{j}^2$, and j_z, i.e., $\mathbf{s} \cdot \mathbf{l}$ is diagonal in this representation. In the transformation from the m_l, m_s representation to the j, m_j representation

$$v_{n, l, j, m_j} = \sum_{m_l, m_s} u_{n, l, m_l, m_s} c_{m_l, m_s, j, m_j} \tag{4.53}$$

or in Dirac notation

$$|n, l, j, m_j\rangle = \sum_{m_l, m_s} |n, l, m_l, m_s\rangle \langle l, s, m_l, m_s | l, s, j, m_j\rangle \tag{4.54}$$

the coefficients $\langle l, s, m_l, m_s | l, s, j, m_j\rangle$ are called Clebsch–Gordan coefficients, and tables of them appear in more advanced books. In taking expectation values, however, we do not need to know what these coefficients are: we only use the fact that $|n, l, j, m_j\rangle$ is an eigenfunction of $\mathbf{l}^2$, $\mathbf{s}^2$, $\mathbf{j}^2$, and j_z. Thus

$$\Delta E_1 = \langle nljm_j | \xi \mathbf{s} \cdot \mathbf{l} | nljm_j \rangle. \tag{4.55}$$

Now

$$\mathbf{j}^2 = (\mathbf{l} + \mathbf{s}) \cdot (\mathbf{l} + \mathbf{s}) = \mathbf{l}^2 + \mathbf{s}^2 + 2\mathbf{s} \cdot \mathbf{l}, \tag{4.56}$$

hence

$$\mathbf{s} \cdot \mathbf{l} = \tfrac{1}{2}\{\mathbf{j}^2 - \mathbf{l}^2 - \mathbf{s}^2\}. \tag{4.57}$$

Finally

$$\Delta E_1 = \frac{\xi}{2} \langle nljm_j | \mathbf{j}^2 - \mathbf{l}^2 - \mathbf{s}^2 | nljm_j \rangle$$

$$= \frac{\xi}{2} \{j(j + 1) - l(l + 1) - s(s + 1)\}. \tag{4.58}$$

This is equivalent to the expression (4.28) for $\langle \mathbf{s} \cdot \mathbf{l} \rangle$ which was used in evaluating $\Delta E''$. Each level, labelled by j, is still $(2j + 1)$-fold degenerate with respect to m_j. Such a degeneracy could be lifted by the application of an external electric or magnetic field, an effect which we are not discussing in this chapter.

We have gone into this derivation in some detail because similar problems arise again and again in the central-field approximation.

4.4. The Lamb shift

The Dirac scheme for the fine structure of hydrogen as illustrated in Fig. 4.1 does not quite agree with experiment. Levels with the same j but different l are not in fact quite coincident. The separations involved are extremely small: the largest separation is that between the $^2S_{1/2}$ and $^2P_{1/2}$ levels† for a given n; for $j > \frac{1}{2}$ the separations are quite negligible. Lamb and Retherford‡ measured accurately by a radiofrequency method the interval $2^2P_{1/2}$—$2^2S_{1/2}$ in hydrogen. In this difficult experiment extraordinary care was taken to achieve high precision, and many small corrections including, for example, the effect of unresolved hyperfine structure were taken into account. They found that the $2^2S_{1/2}$ level lies above the $2^2P_{1/2}$ level by an amount $1,057 \cdot 77 \pm 0 \cdot 10$ MHz, or $0 \cdot 035283$ cm^{-1}, which is only about one-tenth of the fine-structure splitting of the $n = 2$ term. This shift of the S-term is called a Lamb shift.

The effect is explained in terms of the theory of quantum electrodynamics: in fact the development of the theory was greatly stimulated by the experimental measurements. Welton§ has given a qualitative explanation of the Lamb shift as follows: a quantized radiation field has a zero-point energy equivalent to a mean-square electric field so that even in a vacuum there are fluctuations in this zero-point radiation field. These fluctuations cause an electron to execute an oscillatory motion and its charge is therefore smeared out. If the electron is bound, as in hydrogen, by a non-uniform electric field it experiences a different potential from that appropriate to its mean position. Hence the atomic energy levels are shifted. The effect is greatest for s-electrons through a modification of $|\psi(0)|^2$. (The same zero-point fluctuations may be regarded as being responsible for the process of spontaneous emission: a transition occurs through emission *induced* by the appropriate Fourier component of the zero point fluctuating field.)

The theory gives $1057 \cdot 888 \pm 0 \cdot 013$ MHz for the $2^2P_{1/2} - 2^2S_{1/2}$ Lamb shift in hydrogen. Agreement with the original experimental value, $1057 \cdot 77 \pm 0 \cdot 10$ MHz, is very good but not perfect. More modern Lamb shift measurements have been made in the $n = 1$, 2, and 3 terms of

† The notation for a level specifies the value of j as a subscript to the letter code for l. The superscript gives the so-called *multiplicity* of the term—here it is the value of $2s + 1$.

‡ W. E. Lamb and R. C. Retherford, *Phys. Rev.* **72**, 241, 1947.

§ T. A. Welton, *Phys. Rev.* **74**, 1157, 1948.

hydrogen and deuterium, in He^+, and in other hydrogen-like ions up to Ar^{17+}. The best value of the five-structure constant itself was formerly taken to be that derived from Lamb's measurements of the Lamb shift and the matching interval $^2S_{1/2} - {}^2P_{3/2}$ in the $n = 2$ term of deuterium. From the sum of these two measured intervals (equivalent to the fine-structure interval $^2P_{1/2} - {}^2P_{3/2}$), α^{-1} was found to be

$$\alpha^{-1} = 137 \cdot 0365 \pm 0 \cdot 0012. \tag{4.59}$$

Precision measurements of α and of Lamb shifts are of enormous importance as tests of quantum electrodynamics, and there are continuing efforts to improve the measurements† and to investigate other methods of determining α accurately.

Quantum electrodynamics also leads to a modification of the Dirac value, $g_s = 2$, for the electron-spin g-factor. The theory gives

$$g_s = 2(1 + \frac{\alpha}{2\pi} - 0 \cdot 328 \frac{\alpha^2}{\pi^2} + \cdots)$$
$$= 2 \times 1 \cdot 0011596$$
$$= 2 \cdot 0023192. \tag{4.60}$$

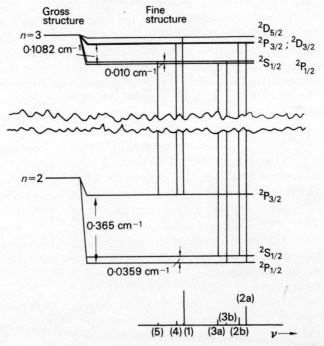

Fig. 4.3. Fine structure of the $n = 2$ and $n = 3$ levels of hydrogen including Lamb shift, with the spectrum of allowed transitions.

† For a review of tests of quantum electrodynamics see, for example, the article by P. J. Mohr in *Atomic Physics*, Vol. 5, p. 37, Plenum Press, 1977.

4.5. Summary of the hydrogen spectrum

Measurements confirm this value, but for many purposes the approximation $g_s = 2.00$ is adequate.

Figure 4.3 shows the fine structure of the $n = 2$ and $n = 3$ levels of hydrogen, modified to take account of the Lamb shift. This structure has also been studied optically by Kuhn and Series.† With great difficulty they were able to resolve the components (3a) and (5), thus obtaining a measurement of the Lamb shift in the terms $n = 2$ and $n = 3$, but nowadays the techniques of laser spectroscopy‡ allow more precise measurements to be made in the optical region.

4.5. Summary of the hydrogen spectrum

Lest the many details which we have discussed so far obscure the general picture of the structure of the energy levels in hydrogen, we now summarize the main theoretical results in tabular form.

GROSS STRUCTURE: the Hamiltonian is $\mathcal{H}_0 = T + V(r)$; $V(r) = -Ze^2/4\pi\varepsilon_0 r$.

Characteristics	*Remarks*
(a) The wave function is separable into radial and angular parts.	Because $V(r)$ is a central field.
(b) The angular constants of the motion are $\mathbf{l}^2$, $\mathbf{s}^2$, l_z, and s_z.	s and l are assumed not to interact with each other.
(c) The energy is a function of n only, and is proportional to Z^2/n^2.	$\langle r^{-1} \rangle$ fixes the magnitude.
There is degeneracy in m_l and m_s;	No axis in space has been physically established.
and in l.	Accidental degeneracy because $V(r)$ has the special form $V \propto r^{-1}$.
(d) The electric dipole selection rules are $\Delta l = \pm 1$;	Change of parity.
$\Delta m_l = 0$, π polarization; $\Delta m_l = \pm 1$, σ polarization.	

FINE STRUCTURE: the Hamiltonian is $\mathcal{H} = \mathcal{H}_0 + \mathcal{H}_1 + \mathcal{H}_2$; $\mathcal{H}_1 = \xi \mathbf{s} \cdot \mathbf{l}$; $\mathcal{H}_2 = $ other relativistic effects.

Characteristics	*Remarks*
(a) $\mathcal{H}_1 = \xi \mathbf{s} \cdot \mathbf{l}$ is treated as a small perturbation.	s and l are now interacting.
(b) The angular constants of the motion are $\mathbf{l}^2$, $\mathbf{s}^2$, $\mathbf{j}^2$, and j_z, *not* l_z, s_z.	There is a torque on s and on l but no torque on j.
(c) The corresponding zeroth-order wave functions in degenerate perturbation theory are $\lvert n, l, j, m_j \rangle$, not $\lvert n, l, m_l, m_s \rangle$.	The perturbation does not mix up states of different j and m_j; $\mathbf{s} \cdot \mathbf{l}$ is diagonal in the $\lvert n, l, j, m_j \rangle$ representation.
(d) The first-order energy shifts are proportional to ξ.	ξ contains the radial part in the form $\langle r^{-3} \rangle$ for $V \propto r^{-1}$.
(e) The vector model pictures a precession of l and s about j.	

† H. G. Kuhn and G. W. Series, *Proc. Roy. Soc.* **A202**, 127, 1950.
‡ See A. Corney, *Atomic and Laser Spectroscopy*, O.U.P., 1977, for a detailed account of modern techniques.

The rate of precession is proportional to ξ.

Hence the rate of precession is linked with the energy splitting.

(f) The fine structure is so small in hydrogen that other relativistic effects ($\mathcal{H}_2$) have to be treated at the same time.

The terms considered are of order v^2/c^2 (or $Z^2\alpha^2$) compared with unity.

(g) The fine structure increases as Z^4 in the iso-electronic sequence.

(h) The fine structure depends on j but not on l for a given n (excluding the Lamb shift).

The accidental degeneracy in l persists because $V(r) \propto r^{-1}$.

(i) The electric dipole selection rule for j is
$$\Delta j = 0, \pm 1.$$

(j) Each level j is $(2j + 1)$−fold degenerate with respect to m_j.

No axis in space has been physically established.

Problems

(Those problems marked with an asterisk are more advanced.)

4.1. In a Stern–Gerlach experiment a well-collimated beam of silver atoms in their ground state ($^2S_{1/2}$) emerges from an oven inside which the atoms are in thermal equilibrium at temperature T. The beam enters a region, of length l, in which there is a strong magnetic field B and a gradient of field $\partial B/\partial z$ perpendicular to the axis of the beam. After leaving this region the beam travels a further distance l' in a field-free region to a detector. Show that in the plane of the detector the deflection s_α of those atoms which had the most probable speed α in the oven is

$$s_\alpha = \pm \frac{\mu_B}{4kT} \frac{\partial B}{\partial z} (l^2 + 2ll'),$$

where μ_B is the Bohr magneton.

Evaluate s_α for $T = 1{,}400$ K, $\partial B/\partial z = 300$ Tm^{-1}, $l = l' = 0 \cdot 1$ m.

*4.2. The electron spin operator **s** can be written in matrix form in terms of the Pauli spin operator $\sigma = 2\mathbf{s}$ where

$$\sigma_x = \begin{pmatrix} 0 & 1 \\ 1 & 0 \end{pmatrix}; \qquad \sigma_y = \begin{pmatrix} 0 & -i \\ +i & 0 \end{pmatrix}; \qquad \sigma_z = \begin{pmatrix} 1 & 0 \\ 0 & -1 \end{pmatrix}.$$

$$\chi(m_s = \tfrac{1}{2}) = \begin{pmatrix} 1 \\ 0 \end{pmatrix}; \qquad \chi(m_s = -\tfrac{1}{2}) = \begin{pmatrix} 0 \\ 1 \end{pmatrix}.$$

(a) Verify that these functions are eigenfunctions of $\mathbf{s}^2$ and s_z, and find the eigenvalues.

(b) Show that not only does **s** satisfy the commutation relations for angular momentum but also

$$\sigma_x^2 = \sigma_y^2 = \sigma_z^2 = 1 \quad \text{and} \quad \sigma_x\sigma_y + \sigma_y\sigma_x = 0 \quad \text{(and cyclically).}$$

(c) Verify the relation

$$(\sigma \cdot \mathbf{F})(\sigma \cdot \mathbf{G}) = \mathbf{F} \cdot \mathbf{G} + i\sigma \cdot \mathbf{F} \times \mathbf{G}$$

where **F** and **G** are any two vector operators which commute with **σ**. Hence show that $(\mathbf{s} \cdot \mathbf{r})^2 = \frac{1}{4} r^2$.

4.3. From the expression

$$-\frac{\hbar^2}{4m_0^2 c^2} e\mathbf{E} \cdot \nabla$$

for the Darwin term in the fine structure of hydrogen, derive by first-order perturbation theory the energy shift

$$\Delta E''' = \frac{\pi \hbar^2}{2m_0^2 c^2} \frac{Ze^2}{4\pi\varepsilon_0} |\psi(0)|^2,$$

which is eq. (4.30). Construct an energy level diagram, drawn to scale, to show separately the contributions $\Delta E'$, $\Delta E''$, and $\Delta E'''$ to the energy shift ΔE (eq. (4.32)) for the $n = 2$ level of hydrogen.

4.4. Estimate the Doppler width of the H_α line, and compare it with the separations of the fine-structure components of the line (a) at 300 K, (b) at 20 K.

4.5. Estimate the electrostatic potential energy of an electron at the surface of a proton and compare it with the rest energy of the electron.

5. Two-electron system

So far we have treated the case of one electron in a potential field. In discussing the fine structure of hydrogen we quoted effective terms in the Hamiltonian of Schrödinger's equation to describe an approximation, up to order v^2/c^2, to the relativistic effects which are properly treated by the Dirac equation. The Dirac equation for one electron in a Coulomb field can be solved exactly, but it is not known how to extend the exact treatment to the case of more than one electron.

As for the wave function we used first of all a spatial function; and when we came to consider spin-dependent terms in the Hamiltonian we built up a simple product of space and spin functions, eq. (4.37). We did this on the grounds that in zeroth order, i.e., in a non-relativistic approximation, there is no interaction between space and spin variables. It then turned out that, because of degeneracy with respect to m_l and m_s, we had to use a particular linear combination of such product functions to treat the spin–orbit interaction in first order of perturbation theory.

In discussing the helium atom we shall now find that we must attach considerable importance to the way in which we build up the wave function to describe the system of a nucleus plus two electrons. Before we come to that we shall delineate the problem by discussing an approximate form for the Hamiltonian.

For two electrons in a non-relativistic approximation we follow the procedure of writing down a Schrödinger equation in which the Hamiltonian contains all the interaction terms we think might be important. These potential energy terms are written in a semi-classical form (an application of the correspondence principle). The problem is still extremely formidable, and a gross simplification is called for: this is achieved by omitting what are believed to be small terms. To clear the air, we shall now list the largest of the terms which we are going to leave out.

If we give the labels 1 and 2 to the electrons whose distances from the nucleus are r_1 and r_2 respectively we shall omit the following effective terms from the Hamiltonian:

    (a) the spin–orbit interaction for electrons 1 and 2: $\xi_1 \mathbf{l}_1 \cdot \mathbf{s}_1 + \xi_2 \mathbf{l}_2 \cdot \mathbf{s}_2$;

    (b) The so-called spin–other-orbit interaction: $\xi_1' \mathbf{l}_1 \cdot \mathbf{s}_2 + \xi_2' \mathbf{l}_2 \cdot \mathbf{s}_1$;

(c) the spin–spin interaction, which has the classical form of a magnetic interaction between two magnetic moments associated with electron spin separated by a distance r_{12}:

$$\zeta'' \left\{ \frac{\mathbf{s}_1 \cdot \mathbf{s}_2}{r_{12}^3} - \frac{3(\mathbf{s}_1 \cdot \mathbf{r}_{12})(\mathbf{s}_2 \cdot \mathbf{r}_{12})}{r_{12}^5} \right\};$$

(d) the magnetic orbit–orbit interaction: $\zeta''' \mathbf{l}_1 \cdot \mathbf{l}_2$.

All the above terms are relativistic in origin and we have left only electrostatic effects in what can be called a non-relativistic Hamiltonian, which we shall now discuss.

5.1. Electrostatic interaction and exchange degeneracy

For two electrons under the approximation considered above, the Schrödinger equation is

$$\left\{ -\frac{\hbar^2}{2m} \nabla_1^2 - \frac{\hbar^2}{2m} \nabla_2^2 - \frac{Ze^2}{4\pi\varepsilon_0 r_1} - \frac{Ze^2}{4\pi\varepsilon_0 r_2} + \frac{e^2}{4\pi\varepsilon_0 r_{12}} \right\} \psi = E\psi. \tag{5.1}$$

or

$$(\mathscr{H}_1 + \mathscr{H}_2 + \mathscr{H}')\psi = E\psi, \tag{5.2}$$

where

$$\mathscr{H}_1 = -\frac{\hbar^2}{2m} \nabla_1^2 - \frac{Ze^2}{4\pi\varepsilon_0 r_1}. \tag{5.3}$$

and similarly for electron 2, with

$$\mathscr{H}' = e^2/4\pi\varepsilon_0 r_{12}. \tag{5.4}$$

ψ is the wave function for the whole system.

The terms (5.3) describe just the motion of each electron separately in the electrostatic field of the nucleus, and eq. (5.4) describes the electrostatic repulsion between the two electrons separated by a distance r_{12}. In the discussion of eq. (5.1), for which there are no spin-dependent terms in the Hamiltonian, we shall leave out all consideration of electron spin even in the wave function. That is, we shall assume that the wave function depends only on the space co-ordinates of the electrons relative to the nucleus. The omission of spin at this stage in no way falsifies the results we are about to obtain for the problem of the electrostatic interaction between two electrons with exchange degeneracy. It is somewhat simpler to analyse the problem without carrying spin wave functions all the way through. In section 5.4 we shall re-introduce the spin wave function and discuss its importance. The reader may then wish to consider how the discussion would have proceeded if we had included spin in the first place.

We attempt a solution of eq. (5.1) by treating $\mathscr{H}'$ as a small perturbation: at least, this is the formal procedure we follow even though, as we shall see, $\mathcal{H}'$ is not necessarily very small. In the zeroth approximation, then, eq. (5.2) becomes

$$(\mathcal{H}_1 + \mathcal{H}_2)\psi_0 = E\psi_0. \tag{5.5}$$

In this approximation electrons 1 and 2 do not interact and we achieve separability of the wave function ψ_0. Since $\mathcal{H}_1$ and $\mathcal{H}_2$ from eq. (5.3) describe just a hydrogen problem for each electron separately, we can write

$$\psi_0 = u_{nlm_l}(\mathbf{r}_1)\, u_{n'l'm_{l'}}(\mathbf{r}_2), \tag{5.6}$$

where

$$\mathcal{H}_1 u_{nlm_l}(\mathbf{r}_1) = E_n u_{nlm_l}(\mathbf{r}_1) \tag{5.7}$$

and

$$\mathcal{H}_2 u_{n'l'm_{l'}}(\mathbf{r}_2) = E_{n'} u_{n'l'm_{l'}}(\mathbf{r}_2), \tag{5.8}$$

with

$$E_0 = E_n + E_{n'}. \tag{5.9}$$

In eqs. (5.7) and (5.8) E_n and $E_{n'}$ are hydrogenic eigenvalues of energy, and their sum is the eigenvalue of eq. (5.5). We now abbreviate the notation: let the label a stand for the set of quantum numbers n, l, m_l, and b for the set n', l', $m_{l'}$, with $a \neq b$. Then eq. (5.6) reads

$$\psi_0 = u_a(1)\, u_b(2) \tag{5.10}$$

with

$$E_0 = E_a + E_b. \tag{5.11}$$

We have labelled the two electrons 1 and 2, meaning that electron 1 has the co-ordinates r_1, θ_1, ϕ_1 in a co-ordinate system whose origin is the nucleus, and similarly for electron 2. But these two electrons are identical in the context of the particular Hamiltonian we are using, that is, they have the same charge and mass. This property presents no difficulties if the electrons are so far apart that they can be thought of as classical particles. We could label them any time we please by virtue of their positions in space, that is, by measuring their positions subject to the limitations of the Uncertainty Principle. The classical situation is the limiting case in which the spatial distribution of each electron charge is, and continues to be, well localized relative to the separation of the electrons. The Hamiltonian in eq. (5.1) is invariant under an exchange of labels 1 and 2 as it must be both classically and quantum mechanically for identical electrons. Yet for the classical situation we could say that the electron which has position vector $\mathbf{r}_1$ is in state a in the central field (eq. (5.7)) while the other electron is in state b (eq. (5.8)). Then the state

76

5.1. Electrostatic interaction and exchange degeneracy

of the combined system of two independent electrons in a central field is expressed as the product $u_a(1)\,u_b(2)$ (eq. (5.10)). However, in an actual helium atom the electrons are sufficiently close together that their charge distributions overlap for much of the time. In this situation the identity of the particles leads to a loss of distinction between the two labels r_1 and r_2 in the sense that the uncertainty principle takes on an over-riding importance in any attempt to measure the positions of the electrons. We must therefore consider another solution to eq. (5.5) with the same energy as the solution (5.10):

$$\psi_0 = u_b(1)\,u_a(2) \tag{5.12}$$

with

$$E_0 = E_b + E_a. \tag{5.13}$$

The existence of two functions $u_a(1)u_b(2)$ and $u_b(1)u_a(2)$, differing only in an exchange of the electron labels and having the same energy, is a case of degeneracy: so-called *exchange degeneracy*. We say that this system is degenerate with respect to exchange of electron labels. We are led to conclude that the use of the two simple product functions (5.10) and (5.12) to describe the helium problem implies far too classical a view of the matter because the two states of the system are actually indistinguishable with respect to these labels. In other words this representation by product functions is a poor one for describing indistinguishable states of overlapping charge clouds; rather, we need a new representation which takes account of indistinguishability. This new representation will be expressed in terms of linear combinations of the old product functions; and it turns out that the particular linear combinations which are needed will appear in a treatment of $\mathcal{H}' = e^2/4\pi\varepsilon_0 r_{12}$ as a perturbation which lifts the exchange degeneracy.

The infinite square matrix of the perturbation $\mathcal{H}'$ has, in general, matrix elements connecting all pairs of states of the same parity, since $\mathcal{H}'$ is an even-parity operator. But most of these pairs of states are not degenerate with each other and the matrix elements of $\mathcal{H}'$ between them do not contribute to an energy shift in a first-order perturbation treatment (they contribute only in higher order, and we shall be able to neglect them). The remaining sets of pairs of degenerate states, $u_a(1)\,u_b(2)$ and $u_b(1)u_a(2)$, may be used to label the rows and columns of separate 2×2 sub-matrices of $\mathcal{H}'$ arranged along the diagonal of the complete matrix. In this section we shall diagonalize one of these sub-matrices of the perturbation $\mathcal{H}'$ to find the energy shifts and wave functions, and we shall discuss the physical consequences afterwards. [This is in contrast to the method of section 4.3 in which we found on physical grounds the constants of the motion, j and m_j, with which to label a representation $(nljm_j)$ in which the perturbation $\mathbf{s} \cdot \mathbf{l}$ was diagonal. In that case we did not find the coefficients in the expansion of the new functions $|nljm_j\rangle$ in terms of the old $|nlm_lm_s\rangle$, nor were we interested in doing so at the time.]

For convenience in what follows we abbreviate the notation further by writing

$$u_a(1)\, u_b(2) = u_{ab}$$
$$u_b(1)\, u_a(2) = u_{ba}$$

(5.14)

in which the label a or b for the quantum state occupied by electron 1 is always written first. With the wave functions u_{ab} and u_{ba} the sub-matrix of $\mathcal{H}' = e^2/4\pi\varepsilon_0 r_{12}$ has four elements:

diagonal:
$$\mathcal{H}'_{11} = \int u_{ab}^* \mathcal{H}' u_{ab}\, d\tau,$$
(5.15)

$$\mathcal{H}'_{22} = \int u_{ba}^* \mathcal{H}' u_{ba}\, d\tau;$$
(5.16)

off-diagonal:
$$\mathcal{H}'_{12} = \int u_{ab}^* \mathcal{H}' u_{ba}\, d\tau,$$
(5.17)

$$\mathcal{H}'_{21} = \int u_{ba}^* \mathcal{H}' u_{ab}\, d\tau.$$
(5.18)

These elements form a 2×2 sub-matrix

$$\begin{bmatrix} \mathcal{H}'_{11} & \mathcal{H}'_{12} \\ \mathcal{H}'_{21} & \mathcal{H}'_{22} \end{bmatrix}.$$

Now $\mathcal{H}'_{11}$ can be written

$$\mathcal{H}'_{11} = \int \frac{(-e)|u_a(1)|^2\, d\tau_1\, (-e)|u_b(2)|^2\, d\tau_2}{4\pi\varepsilon_0 r_{12}} = \int \frac{\rho_a(1)\rho_b(2)\, d\tau_1\, d\tau_2}{4\pi\varepsilon_0 r_{12}},$$

(5.19)

similarly

$$\mathcal{H}'_{22} = \int \frac{\rho_b(1)\rho_a(2)\, d\tau_1\, d\tau_2}{4\pi\varepsilon_0 r_{12}}.$$
(5.20)

These integrals represent just a classical repulsion energy for two electrostatic charge distributions of charge densities ρ_a and ρ_b. Interchanging labels 1 and 2 makes no difference to the integral, so

$$\mathcal{H}'_{11} = \mathcal{H}'_{22} = J \text{ (say)}.$$
(5.21)

J is called the *direct integral*. Also

$$\mathcal{H}'_{12} = \mathcal{H}'_{21} = K \text{ (say)}.$$
(5.22)

K is called the *exchange integral*: it is a quantum-mechanical interference term.

5.1. Electrostatic interaction and exchange degeneracy

In order to diagonalize the sub-matrix of $\mathcal{H}'$ we form new zeroth-order wave functions which are linear combinations of the old functions. We write

$$U = c_1 u_{ab} + c_2 u_{ba}, \tag{5.23}$$

with

$$|c_1|^2 + |c_2|^2 = 1 \quad \text{for normalization.} \tag{5.24}$$

The functions U are still (two-fold degenerate) eigenfunctions of the zeroth-order Hamiltonian $\mathcal{H}_0$, with energy E_0, as in eq. (5.5) whatever the values of c_1 and c_2, but now c_1 and c_2 are to be chosen in such a way that $\mathcal{H}'$ also has only diagonal sub-matrix elements in the new representation. This is achieved as follows: in the exact eq. (5.2) we write, as an approximation,

$$\psi = U + \Delta U \tag{5.25}$$

where ΔU is the first-order correction to U brought about by the mixing into U of other states with energy $\neq E_0$ on account of the perturbation $\mathcal{H}'$. With $E = E_0 + \Delta E$, where ΔE is the first-order energy shift, eq. (5.2) becomes, to first order,

$$(\mathcal{H}_0 + \mathcal{H}')U + \mathcal{H}_0 \Delta U = (E_0 + \Delta E)U + E_0 \Delta U. \tag{5.26}$$

Subtracting eq. (5.5) we get

$$\mathcal{H}'U + \mathcal{H}_0 \Delta U = \Delta E U + E_0 \Delta U. \tag{5.27}$$

We now multiply on the left by u_{ab}^* and u_{ba}^* in turn and integrate over all space. Taking account of the fact that the wave functions contained in ΔU are orthogonal to u_{ab} and u_{ba} we obtain the pair of equations

$$c_1 \Delta E = c_1 \mathcal{H}'_{11} + c_2 \mathcal{H}'_{12}, \tag{5.28}$$

$$c_2 \Delta E = c_1 \mathcal{H}'_{21} + c_2 \mathcal{H}'_{22}. \tag{5.29}$$

Equations (5.28) and (5.29) are a pair of simultaneous equations in the coefficients c_1 and c_2. To avoid the trivial solution $c_1 = c_2 = 0$ we require the condition

$$\begin{vmatrix} (\mathcal{H}'_{11} - \Delta E) & \mathcal{H}'_{12} \\ \mathcal{H}'_{21} & (\mathcal{H}'_{22} - \Delta E) \end{vmatrix} = 0. \tag{5.30}$$

Equation (5.30) is called the *secular equation* for the problem. From it are derived two solutions for the energy shift ΔE arising from the perturbation. In the present case, using eq. (5.21) and (5.22), we have

$$\begin{vmatrix} J - \Delta E & K \\ K & J - \Delta E \end{vmatrix} = 0 \tag{5.31}$$

which gives

$$\Delta E = J \pm K. \qquad (5.32)$$

Associated with each energy shift there is a wave function U satisfying eq. (5.27). In anticipation we give them the names U_S associated with $\Delta E = J + K$ and U_A associated with $\Delta E = J - K$. To express U_S and U_A in terms of u_{ab} and u_{ba} we evaluate c_1 and c_2 for the two cases $\Delta E = J \pm K$ from eq. (5.28):

$\Delta E = J + K$;

$$(J + K)c_1 = Jc_1 + Kc_2, \qquad (5.33)$$

whence

$$c_1 = c_2. \qquad (5.34)$$

From eq. (5.24) we find $|c_1|^2 = \tfrac{1}{2}$ and we choose the phase $c_1 = +(2)^{-1/2}$. Thus

$$U_S = (2)^{-1/2} (u_{ab} + u_{ba}). \qquad (5.35)$$

For $\Delta E = J - K$,

$$(J - K)c_1' = Jc_1' + Kc_2', \qquad (5.36)$$

whence

$$c_2' = -c_1'. \qquad (5.37)$$

Again we choose the phase $c_1' = +(2)^{-1/2}$, so $c_2' = -(2)^{-1/2}$ and

$$U_A = (2)^{-1/2} (u_{ab} - u_{ba}). \qquad (5.38)$$

Thus we have found the two functions U_S and U_A which diagonalize the perturbation $e^2/4\pi\varepsilon_0 r_{12}$. (It may easily be verified that

$$\int U_s^* (e^2/4\pi\varepsilon_0 r_{12}) U_A \, d\tau = 0.)$$

U_S remains unchanged on exchange of electron labels 1 and 2, whereas U_A changes sign. U_S is said to be symmetric and U_A anti-symmetric with respect to exchange, hence the choice of symbol for the subscript.

We might have argued from the beginning that a *definite* symmetry under exchange of labels is just the property that is needed for zeroth-order wave functions, degenerate with respect to exchange, which are to describe indistinguishable states. We can introduce an abstract operator which represents exchange of labels. If, more generally, $\psi(r_1 \cdots r_i r_j \cdots)$ is a wave function for a many-electron system, the operator P_{ij} is defined by

$$P_{ij}\psi(r_1 \cdots r_i r_j \cdots) = \psi(r_1 \cdots r_j r_i \cdots). \qquad (5.39)$$

5.2. The ground state of helium

Then

$$P_{ij}^2 \psi(r_1 \cdots r_i r_j \cdots) = P_{ij}\psi(r_1 \cdots r_j r_i \cdots)$$
$$= \psi(r_1 \cdots r_i r_j \cdots) \quad (5.40)$$

and P_{ij}^2 has the eigenvalue 1. Any function is an eigenfunction of P_{ij}^2. But those functions which are eigenfunctions of P_{ij} itself, with eigenvalues ± 1, are the particular ones which have a definite symmetry with respect to exchange. We have found that for two electrons U_S and U_A are the eigenfunctions of P_{12} with eigenvalues $+1$ and -1 respectively:

$$P_{12}U_S = +U_S, \qquad P_{12}U_A = -U_A, \quad (5.41)$$

whereas u_{ab} and u_{ba} do not satisfy such eigenvalue equations. The importance of this symmetry classification lies in the fact that P_{12} commutes with the total Hamiltonian of eq. (5.2): therefore P_{12} represents a *constant of the motion*. It is the very operator whose eigenfunctions describe satisfactorily a degenerate system of indistinguishable states and which at the same time diagonalize the perturbation $e^2/4\pi\varepsilon_0 r_{12}$ and hence solve the two-electron problem in perturbation theory when exchange degeneracy is present. This is analogous to the previous example of section 4.3, that of diagonalizing the matrix of $\mathbf{s} \cdot \mathbf{l}$: $\mathbf{j}^2$ and j_z, representing constants of the motion, provided the labels j and m_j for the new representation $(lsjm_j)$ which diagonalized $\mathbf{s} \cdot \mathbf{l}$.

A state of definite exchange symmetry keeps that symmetry in time. Even when the Hamiltonian is extended to include interaction with a radiation field one cannot have electromagnetic transitions between symmetric and anti-symmetric states because the multipole-moment operators are symmetrical with respect to exchange. For example, an electric dipole moment matrix element would be $\int U_S^*(e\mathbf{r}_1 + e\mathbf{r}_2)U_A \, d\tau$.

This whole matrix element is anti-symmetric with respect to exchange, and since exchange of labels cannot affect the value of the integral we must have

$$\int U_S^*(e\mathbf{r}_1 + e\mathbf{r}_2)U_A \, d\tau = -\int U_S^*(e\mathbf{r}_2 + e\mathbf{r}_1)U_A \, d\tau = 0. \quad (5.42)$$

Hence we have the rule that electromagnetic transitions between symmetric and anti-symmetric states in co-ordinate space are forbidden (in the approximation that interactions between electron spin and orbital motion are neglected).

5.2. The ground state of helium

Before we discuss the energy spectrum of the two-electron system in general from eq. (5.32), we shall examine the ground state of helium.

In the zeroth approximation, eq. (5.5), the electrons do not interact and the energy is given by eq. (5.9) with the electrons labelled by $n_1 = 1$, $l_1 = 0$, $m_{l_1} = 0$, $n_2 = 1$, $l_2 = 0$, $m_{l_2} = 0$. This corresponds to a hydrogenic ground state for each single-electron problem separately. The configuration is written $1s^2$. The labels a and b of the last section are therefore equal, and from eq. (5.38)

$$U_A \text{ vanishes.}$$

We have only the symmetric function $u_a(1)u_a(2) = u_{1s}(1)u_{1s}(2)$ which is non-degenerate with respect to exchange. The unperturbed energy is

$$E(1s^2) = 2E(1s) \tag{5.43}$$

where, since $E_n \propto Z^2$ and $Z = 2$, $E(1s) = -4$ Rydbergs ≈ -54.4 eV. Hence $E(1s^2) \approx -108.8$ eV. That is, the work required to strip two electrons from neutral helium in its unperturbed ground state is just twice the ionization potential of He$^+$, which is a hydrogenic ion with $Z = 2$ (see Fig. 5.1).

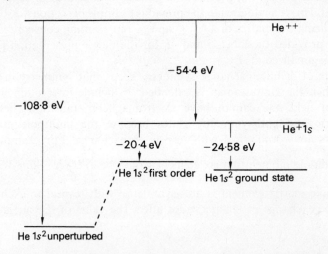

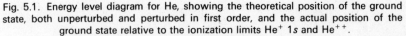

Fig. 5.1. Energy level diagram for He, showing the theoretical position of the ground state, both unperturbed and perturbed in first order, and the actual position of the ground state relative to the ionization limits He$^+$ $1s$ and He^{++}.

The term $e^2/4\pi\varepsilon_0 r_{12}$ can be treated by non-degenerate perturbation theory since the unperturbed state $u_{1s}(1)u_{1s}(2)$ is non-degenerate with respect to l_1, m_{l_1}, l_2 and m_{l_2}. In first order the energy shift is a direct integral of the form of eq. (5.19) for which the result is (see problem 5.3)

$$\Delta E(1s^2) = \langle e^2/4\pi\varepsilon_0 r_{12} \rangle = \frac{1}{4\pi\varepsilon_0} \frac{5}{4} Z \frac{e^2}{2a_0} \approx 34 \text{ eV}. \tag{5.44}$$

5.3. The excited states of helium

This result would make the ionization potential of neutral helium $54.4 - 34 = 20.4$ eV. A variational procedure gives a better result (see problem 5.2). Much more exact numerical calculations give agreement with the experimental value of 24.580 eV. The discrepancy of 14 per cent between the first-order perturbation result and the correct ionization potential is not surprising since the perturbation can hardly be regarded as small (an energy shift of 34 eV out of a total energy of 108.8 eV). These results are summarized in Fig. 5.1.

5.3. The excited states of helium

From eqs. (5.11) and (5.32) we have the energy of an excited state of helium measured from the ionization limit of He^1, i.e., from the state of the He^{++} ion:

$$E = E_{n_1} + E_{n_2} + J \pm K. \qquad (5.45)$$

Since excitation of two electrons is rare, and in helium would give energy levels above the ionization limit of the neutral atom, in order to discuss the discrete levels we consider one of the electrons to be labelled by $1s$ and we subtract from eq. (5.45) the energy E_{1s}. Then for the configuration $1snl$ we have the energy

$$E = E_n + J \pm K \qquad (5.46)$$

where now the energy is measured relative to the ionization limit of the neutral atom (the ground state of He^+). In this lowest approximation† the unperturbed energy E_n can be regarded as being (a) shifted by the direct integral

$$J(1s; nl) = \langle 1s; nl|e^2/4\pi\varepsilon_0 r_{12}|1s; nl\rangle, \qquad (5.47)$$

and (b) split by the exchange integral

$$K(1s; nl) = \langle 1s; nl| e^2/4\pi\varepsilon_0 r_{12} |nl; 1s\rangle. \qquad (5.48)$$

This is illustrated in Fig. 5.2. The upper and lower signs in eq. (5.46) belong to the symmetric and anti-symmetric states respectively.

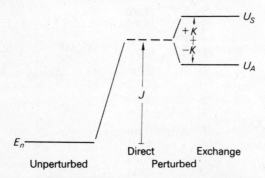

Fig. 5.2. Schematic energy level diagram for an excited state of He, showing the effect of the direct and exchange integrals.

† See problem 5.5.

The effect of the direct integral is to cancel out part of the central attraction of the nuclear charge. Another way of looking at this is in terms of *screening*. Since the $1s$ electron has a charge distribution which is on the average close to the nucleus (we therefore call the $1s$ electron the inner electron), and the nl electron (the outer electron) has its charge distribution further out, the inner electron screens the nuclear charge from the outer electron. This screening is more and more complete for large n and l, and in this case it would be a good approximation to rewrite the potential energy terms in the Hamiltonian, not with the mutual interaction $e^2/4\pi\varepsilon_0 r_{12}$, but in the effective form

$$V = Ze^2/4\pi\varepsilon_0 r_{1s} - (Z - 1)e^2/4\pi\varepsilon_0 r_{nl} \qquad (5.49)$$

which expresses the fact that the inner electron sees the full nuclear charge Ze and the outer electron sees an effectively screened nuclear charge $(Z - 1)e$. According to this approximation the energy levels of neutral helium would be hydrogen-like (for large n and l) with an effective nuclear charge of 1. One can see how close this approximation is from the data of table 5.1.

The direct coulomb interaction $e^2/4\pi\varepsilon_0 r_{12}$ raises the degeneracy in l

Table 5.1. Energy levels in helium

Configuration	Term	Helium $-E$ cm^{-1}	singlet–triplet difference, cm^{-1}	hydrogenic $-E$ cm^{-1}
$1s^2$	^{3}S	missing		109,678
	^{1}S	198,311		
$1s2s$	^{3}S	38,461	6,422	
	^{1}S	32,039		
				27,419
$1s2p$	^{3}P	29,230	2,048	
	^{1}P	27,182		
$1s3s$	^{3}S	15,080	1,628	
	^{1}S	13,452		
$1s3p$	^{3}P	12,752	645	
	^{1}P	12,107		12,186
$1s3d$	^{3}D	12,215	3	
	^{1}D	12,212		
$1s4s$	^{3}S	8,019	643	
	^{1}S	7,376		
$1s4p$	^{3}P	7,100	276	
	^{1}P	6,824		
				6,854
$1s4d$	^{3}D	6,872	2	
	^{1}D	6,870		
$1s4f$	^{3}F	6,864·4	0·6	
	^{1}F	6,863·8		

5.3. The excited states of helium

because the amount of screening depends on the eccentricity of the orbit of the outer electron. For example, for $n = 2$ in helium the $1s2p$ configuration lies about 7,000 cm^{-1} above the $1s2s$ configuration instead of being degenerate with it (see Fig. 5.3 and table 5.1).

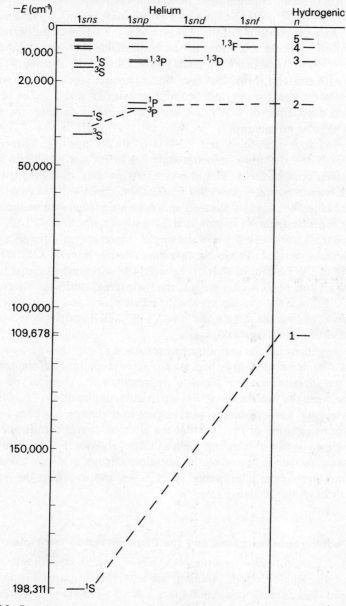

Fig. 5.3. Energy level diagram for helium. For comparison the hydrogenic levels are shown on the right.

We shall go into the question of the constants of the motion in general for a two-electron system later (chapter 7). For the moment we want to clarify the labelling of the terms in helium. The total orbital angular momentum $\mathbf{L}$ formed by the vector addition $\mathbf{L} = \mathbf{l}_1 + \mathbf{l}_2$ becomes $\mathbf{L} = \mathbf{l}$ for the $1snl$ configuration of helium because $l_1 = 0$ and $l_2 = l$. Even in the more general case of a configuration $n_1 l_1$, $n_2 l_2$ $\mathbf{L}^2$ and L_z would represent constants of the motion under the action of the perturbation $e^2/4\pi\varepsilon_0 r_{12}$ because this interaction is internal to the atom and there is no torque on the total orbital angular momentum (in the absence of interaction with electron spin): that is $e^2/4\pi\varepsilon_0 r_{12}$ commutes with $\mathbf{L}^2$ and L_z. The exchange operator P_{12} also commutes with $\mathbf{L}^2$ and L_z. The terms, now non-degenerate in $L = l$, are labelled by the code letter for total orbital angular momentum.

The exchange integral K (eq. (5.48)) is always positive. Therefore, from eq. (5.46), the anti-symmetric state lies below the symmetric state for a given configuration. The anti-symmetric states are called triplets and the symmetric states are called singlets, a nomenclature having to do with the eigenfunctions of electron spin which are discussed in the next section. This designation is indicated by a superscript, e.g., ^3P, ^1P. It is very apparent from table 5.1 how the singlet–triplet energy difference falls off as n and l increase. This is because the exchange integral $K(1s;nl)$ is a measure of the amount of overlap between the unperturbed radial functions $R(1s)$ and $R(nl)$, which are just the hydrogenic functions illustrated in Fig. 2.1. Since the overlap between the $1s$ function and the nl function becomes negligible for large n and l, eq. (5.49) well describes the classical limit of screening without exchange.

The way in which the exchange phenomenon affects the energy levels is hard to understand at first, because the actual term in the Hamiltonian which we are considering $e^2/4\pi\varepsilon_0 r_{12}$, represents just a classical repulsive force between the two electrons. But a particular choice of wave functions, symmetric and anti-symmetric with respect to exchange, has been forced on us in recognition of the fact that the electron charge clouds overlap, although we persist in labelling them as if they did not. It is in the use of these wave functions to find the expectation value of $e^2/4\pi\varepsilon_0 r_{12}$ that the quantum-mechanical interference effect arises and so affects the energy levels through the exchange integral.

5.4. Electron spin functions and the Pauli exclusion principle

Provided there is no interaction between spin and co-ordinates, the zeroth-order single-electron functions are separable in space co-ordinates and spin. We can write combined space and spin functions as a product

$$\psi(1) = u_{nlm_l}(1)\chi_{\pm}(1) \tag{5.50}$$

5.4. Electron spin functions and the Pauli exclusion principle

where the label (1) refers to electron number one. $\chi_{\pm}(1)$ are the eigenfunctions of s_1^2 and s_{1_z} with eigenvalues $s_1(s_1 + 1) = \frac{3}{4}$ and $m_{s_1} = \pm \frac{1}{2}$.

For two electrons we have in this representation products of spin functions $\chi_{\pm}(1)\chi_{\pm}(2)$, of which there are four possibilities corresponding to the four possible combinations of $m_{s_1} = \pm \frac{1}{2}$, $m_{s_2} = \pm \frac{1}{2}$. These product functions do not have a definite symmetry with respect to exchange. We can adopt a new representation in which the functions are simultaneous eigenfunctions of the square of the total spin angular momentum $S^2 = (s_1 + s_2)^2$, with eigenvalue $S(S + 1)$, and of its projection $S_z = s_{1_z} + s_{2_z}$, with eigenvalue M_S. At the same time these new functions have definite symmetry with respect to exchange. They are normalized linear combinations of the old product functions:

$$\chi_S = \begin{cases} \chi_+(1)\chi_+(2), \\ (2)^{-1/2}\{\chi_+(1)\chi_-(2) + \chi_-(1)\chi_+(2)\}, \\ \chi_-(1)\chi_-(2); \end{cases} \tag{5.51}$$

and

$$\chi_A = (2)^{-1/2}\{\chi_+(1)\chi_-(2) - \chi_-(1)\chi_+(2)\}. \tag{5.52}$$

For the three functions χ_S of eq. (5.51) which are symmetric with respect to exchange one can verify that $S = 1$ and $M_S = 1, 0, -1$. These functions form a triplet, triply degenerate with respect to M_S in the absence of a physically established axis in space. The function χ_A of eq. (5.52) is a singlet function, anti-symmetric with respect to exchange, and corresponding to $S = 0$, $M_S = 0$. (The operators S^2 and S_z commute with the Hamiltonian of eq. (5.2) and also with the exchange operator P_{12}, which accounts for the fact that S^2 and S_z represent constants of the motion and that their eigenfunctions also have a definite exchange symmetry.)

Now we can combine the two spin functions χ_S and χ_A with the two co-ordinate functions U_S and U_A of eqs. (5.35) and (5.38) in a representation $(LM_L; SM_S)$. The Hamiltonian of eq. (5.2) which describes the electrostatic interactions does not operate on the spin functions, and so electron spin does not enter directly in a description of the structure of electrostatic energy levels which we have discussed so far in this chapter.

But when we consider the combined functions $U_S\chi_S$, $U_A\chi_A$, and $U_S\chi_A$, $U_A\chi_S$ a new result of extreme importance emerges. The first pair of functions

$$\psi_S = \begin{matrix} U_S\chi_S \\ U_A\chi_A \end{matrix} \tag{5.53}$$

are totally symmetric with respect to exchange in co-ordinates *and* spin while the second pair

$$\psi_A = \frac{U_S \chi_A}{U_A \chi_S} \tag{5.54}$$

are totally anti-symmetric. In section 5.1 we stated the relatively weak rule that electromagnetic transitions between states U_S and U_A are forbidden—weak, because of the qualification that spin-dependent interactions were neglected. Now we state the rule, due originally to Heisenberg and Dirac, that transitions between totally symmetric and totally anti-symmetric states ψ_S and ψ_A are *not allowed for any process whatever*. In fact it appears that there are two kinds of particles: bosons, which have integral spin and are described by totally symmetric functions of the kind ψ_S; and fermions, which have half-integral spin and are described by totally anti-symmetric functions of the kind ψ_A. Different kinds of quantum statistics are used to describe assemblies of the two kinds of particles.

Confining our discussion to electrons, which are fermions, we find that one way of expressing the *Pauli exclusion principle* is that *for electrons, only functions which are totally anti-symmetric with respect to exchange in co-ordinates and spin are allowed*. Thus for two electrons the functions U_S, symmetric in co-ordinate space, are associated only with χ_A and are therefore *singlets*, while the functions U_A, anti-symmetric in co-ordinate space, are *triplets*. We are now in a position to see what quantum statistics has to say about the non-relativistic helium problem. It is simply this: that both sets of states $U_S \chi_A$ and $U_A \chi_S$ *actually exist*. If, say, we had been dealing with the hypothetical case in which the two valence particles were identical bosons of spin zero, the calculation of energies including exchange integrals would have proceeded as in section 5.3, but those levels associated with U_A would not have existed.

We mention here, in passing, a fairly obvious selection rule for the total spin quantum number S. When there is assumed to be no interaction between spin and orbit S is a good quantum number. The operators for electric multipole moments do not operate on the spin functions, so in the coupled representation $(LM_L; SM_S)$ matrix elements of electric multipole moments taken between spin functions of different S, i.e., between χ_A and χ_S, vanish because the spin functions χ_A and χ_S are orthogonal. Therefore

$$\Delta S = 0. \tag{5.55}$$

This rule forbids transitions between the singlet and triplet terms in helium, but in this case it adds nothing new because under the same approximation it corresponds to the rule $U_S \nrightarrow U_A$ in co-ordinate space. We have already derived the latter rule, eq. (5.42), from considerations of exchange symmetry in co-ordinate space alone when there is no interaction between spin and orbit. Also from consideration of exchange symmetry in co-ordinate space we found (eq. (5.38)) that the term $1s^2\ {}^3S$ is missing. The triplet term of lowest energy is $1s2s\ {}^3S$ (see Fig. 5.3) and this term is *metastable*, that is, it is an excited state but it cannot decay to a lower energy state by radiation. In the present approximation it has an infinite lifetime like the ground state. (The singlet term $1s2s^1S$ is also metastable, but for different reasons not connected with exchange symmetry. We shall discuss selection rules in general later.) The two systems of energy levels in helium, singlet and triplet, corresponding to symmetric and anti-symmetric co-ordinate states U_S and U_A respectively, used to be called para- and ortho-helium. To the extent that radiative transitions between them are forbidden they can be regarded as quite separate. The fact that the ground state belongs to the singlet system but that both systems are excited in a discharge tube is due to collisions involving spin-dependent forces.

Spin-dependent terms in the Hamiltonian give rise to fine structure in the triplet terms of helium. We shall not discuss this now, but shall return to it later (chapter 7).

We have discussed the character of the *energy levels* in a two-electron system, under electrostatic forces only, by considering exchange symmetry in co-ordinate space only, not in spin space. Alternatively, one could state the Pauli exclusion principle (which includes spin) as a fundamental principle, and arrive at the conclusion that the same two co-ordinate functions U_S and U_A are the ones which are required to describe the energy levels. The exclusion principle is needed to establish the *existence* of the levels associated with U_S and U_A, and in the case of three or more electrons it is vital for determining the allowed states of the system. We have treated the two-electron system in the manner already presented in this chapter in order to bring out clearly the dependence of the energy on *electrostatic* integrals.

5.5. The periodic system

In general, for N non-interacting electrons with no spin–orbit interaction we find it convenient to go back to the single-particle representation in which n, l, m_l and m_s are specified for each electron. The wave functions which are products of co-ordinate and spin functions (eq. (5.50)) are called single-particle functions or spin-orbitals. We shall now write them

$$\psi_\alpha(1) = u_{nlm_l}(1)\chi_\pm(1) = u_a(1)\chi_\pm(1) \qquad (5.56)$$

where the subscript α stands for the set of four quantum numbers n, l, m_l, m_s. In this representation the exclusion principle may be stated in its original form: no two electrons in an atom may be labelled by the same set of four quantum numbers n, l, m_l, m_s. In terms of the co-ordinate labels only, one may say: not more than two electrons can have the same set of three quantum numbers n, l, m_l (then two electrons which have the same

n, l, m_l must have different m_s, $+\frac{1}{2}$ and $-\frac{1}{2}$). Electrons which have the same n, l are called *equivalent* electrons. Thus not more than $2(2l + 1)$ electrons can be equivalent to each other because for a given l there are $2l + 1$ possible values of m_l, with two possible values of m_s. The relation between Pauli's exclusion principle and the more general symmetry principle of Heisenberg and Dirac is as follows: we require a normalized linear combination of products of single-particle functions $\psi_\alpha(1)\psi_\beta(2) \cdots \psi_\nu(N)$ for N non-interacting electrons constructed in such a way that the total function is totally anti-symmetric with respect to exchange of electrons. Slater showed that a determinant satisfies this requirement:

$$\psi_A = (N!)^{-1/2} \begin{vmatrix} \psi_\alpha(1) & \psi_\alpha(2) & \cdots & \psi_\alpha(N) \\ \psi_\beta(1) & \psi_\beta(2) & \cdots & \\ \vdots & \vdots & & \\ & & & \psi_\nu(N) \end{vmatrix}$$

(5.57)

This form guarantees that $\psi_A = 0$ if any pair of sets of four single-electron quantum numbers are equal, e.g., if $\alpha = \beta$.

For two electrons eq. (5.57) becomes

$$\psi_A = (2)^{-1/2} \begin{vmatrix} \psi_\alpha(1) & \psi_\alpha(2) \\ \psi_\beta(1) & \psi_\beta(2) \end{vmatrix} = (2)^{-1/2}\{\psi_\alpha(1)\psi_\beta(2) - \psi_\beta(1)\psi_\alpha(2)\}.$$

(5.58)

Of course, since we are back in the n, l, m_l, m_s representation it is not necessarily true that eq. (5.58) as it stands is an eigenfunction of $\mathbf{S}^2$ and S_z and of $\mathbf{L}^2$ and L_z in the coupled representation LM_LSM_S in which, for example, we described the helium problem with $\mathbf{s}_1 + \mathbf{s}_2 = \mathbf{S}$ and $\mathbf{l}_1 = 0$, $\mathbf{l}_2 = \mathbf{L}$. We should have to construct linear combinations of determinantal product functions to achieve this. For example, the particular linear combination which gives $U_A\chi_S$ where

$$U_A = (2)^{-1/2}\{u_a(1)u_b(2) - u_b(1)u_a(2)\}$$

and $\chi_S = (2)^{-1/2}\{\chi_+(1)\chi_-(2) + \chi_-(1)\chi_+(2)\}$ corresponding to $S = 1$, $M_S = 0$ from eq. (5.51) is

$$U_A\chi_S(S = 1, M_S = 0) = (2)^{-1/2}\left\{(2)^{-1/2}\begin{vmatrix} u_a(1)\chi_+(1) & u_a(2)\chi_+(2) \\ u_b(1)\chi_-(1) & u_b(2)\chi_-(2) \end{vmatrix}\right.$$

$$\left. + (2)^{-1/2}\begin{vmatrix} u_a(1)\chi_-(1) & u_a(2)\chi_-(2) \\ u_b(1)\chi_+(1) & u_b(2)\chi_+(2) \end{vmatrix}\right\}$$

(5.59)

This can easily be verified by writing out the determinants and rearranging the terms in the sum. However, we shall not have occasion now to go into complicated details of this kind. We give this example to illustrate the importance of realizing which one of various representations is being

5.5. The periodic system

used in the description of a problem, that is to say which variables are being chosen to specify the constants of the motion in a given approximation.

The classification of the elements by shells† of single electrons of given n, l to form a periodic system depends on the exclusion principle. A knowledge of the ordering of the configurations assigned to the ground states of elements as a function of Z depends on detailed calculation of the energy, which involves investigation of radial behaviour. At the beginning of the periodic table the electrons each have the lowest n consistent with the exclusion principle; for a given n they have the lowest l. This is the behaviour expected of weakly interacting electrons.

The labels of the whole array of a determinantal product function such as eq. (5.57) are fixed if just the labels of the diagonal elements are specified. Thus there is a convenient short-hand notation for a configuration of equivalent electrons. For a given n, l the values of m_{l_i} are written out in line and above each is written $+$ or $-$ corresponding to $m_{s_i} = \pm\frac{1}{2}$. For example, one of the many possibilities for three equivalent $4d$ electrons is $\{\overset{+}{2} - \overset{-}{2} - \overset{+}{1}\}$. The normalization factor $(6)^{-1/2}$ can be written in front of the curly bracket, and the relative phase of a wave function is given by the property of a determinant on exchange of rows or columns: for example

$$\{\overset{+}{2} - \overset{-}{2} - \overset{+}{1}\} = -\{-\overset{-}{2}\,\overset{+}{2} - \overset{+}{1}\}. \tag{5.60}$$

For equivalent electrons the exclusion principle states that no two pairs of symbols m_l, m_s may be the same. For helium, $Z = 2$, the two electrons can have $n = 1$ and $l = 0$, that is the configuration $1s^2$ with determinantal state

$$1s^2; \qquad \{\overset{+}{0}\,\overset{-}{0}\}.$$

For the ground state of lithium, $Z = 3$, there is no room for the third electron in the $1s$ shell and it must occupy the next shell, $2s$, which becomes closed at $Z = 4$, beryllium. Beginning at $Z = 5$ the $2p$ shell becomes occupied by the fifth electron: the configuration of the ground state of boron is $1s^22s^22p$. We can proceed with this book-keeping and find how many electrons a p-shell can accommodate. It is best systematically to keep M_S, which is the *algebraic* sum $\sum_i m_{s_i}$, as large as possible, and for a given M_S to maximize $M_L = \sum_i m_{l_i}$. In this way we find that only six electrons can fit into a p-shell, and the wave function is

$$np^6; \qquad \{\overset{+}{1}\,\overset{+}{0} - \overset{+}{1}\,\overset{-}{1}\,\overset{-}{0} - \overset{-}{1}\}.$$

† There is some confusion about the use of the word 'shell'. The word appears in the general expression 'shell-structure'. Sometimes it is used specifically to denote a set of electrons of given n, in which case electrons of given n, l occupy a *sub-shell*. This is also the nomenclature of X-ray spectroscopy: the K-, L-, M-, and N-shells correspond to $n = 1, 2, 3, 4$. We shall use the word to describe electrons of given n, l. The closed-shell configuration of a rare gas marks the end of a *period*.

For a closed p-shell, therefore, the maximum value of M_S equals zero, similarly $M_L(\text{max}) = 0$, corresponding to $S = 0$, $L = 0$ which describes a ^1S term with no resultant orbital or spin angular momentum.

All closed shells give rise to ^1S terms. In particular the rare gases are very stable, for it turns out that an additional electron has to go into the next shell of higher n, forming an alkali atom. In the non-interacting particle approximation this valence electron is much less tightly bound.

The first departure from the ordering according to the lowest value of n occurs at $Z = 19$. The configuration of argon at $Z - 18$ is $1s^2 2s^2 2p^6 3s^2 3p^6$. The $3p$ shell is closed, but the $3d$ shell is still empty. However, the next electron added does not begin the $3d$ shell, with lowest possible n and relatively large l, but goes into the $4s$ shell with larger n but small l. The argon core effectively screens the nuclear charge from a $3d$ electron, but such is the penetration of the core by a $4s$ electron that the latter is more tightly bound and the alkali configuration with a $4s$ valence electron is energetically preferred. The appearance of the $3d$ electron in a ground configuration is deferred until scandium, $Z = 21$, which has the ground configuration $1s^2 2s^2 2p^6 3s^2 3p^6 4s^2 3d$. The way in which the ordering depends on n *and* l has been described by an empirical rule: shells are filled in order of increasing $n + l$, and for a given $n + l$ in order of increasing n.

Configurations of the kind $nd^x(n + 1)s^2$ or $nd^{x+1}(n + 1)s$ occur for $n = 3, 4$, and 5 in the periodic table. As x increases the d-shell is filled up. These three sets of *transition* elements are called the *iron, palladium*, and *platinum* groups respectively. On ionization they tend first to lose their s-electrons, so that as ions they have incomplete d-shells. This is responsible in particular for their paramagnetic behaviour. A similar filling of a so-called 'inner' shell occurs in the *rare-earths* which have configurations of the kind $4f^x 5s^2 5p^6 6s^2$ or sometimes $4f^{x-1} 5s^2 5p^6 5d 6s^2$. Analogous to the rare-earths are the elements, beginning with actinium, in which the $5f$ shell is filling up. In this paragraph we have examples of the competition between different configurations for the ground state of an element. In these cases the competing configurations usually differ in electrons having the same $n + l$. For example, $4f(n + l = 4 + 3 = 7)$ competes with $5d(n + l = 5 + 2 = 7)$. If the one configuration describes the ground state the other will be found as a low-lying excited state.

Indeed, in Ni ($Z = 28$) the term $3d^9 4s$ ^3D lies so close to the ground term $3d^8 4s^2$ ^3F that their separation is less than the fine structure splitting of each, which extends over about $2{,}000$ cm^{-1} (see Fig. 5.4). Furthermore, the two configurations have the same parity so very little meaning can be attached to the phrase 'ground configuration' in this case, for there will be strong mixing of the two configurations through the electrostatic interaction between electrons.

In contrast to these examples of competition for the ground state in which one of the competing electrons has $l \geqslant 2$, the rare gases all have configurations of the type np^6. They end a period in which the np-shell has been filled up in strict order of increasing Z, and their first excited states $np^5(n + 1)s$ as well as their ionization limits lie relatively high in energy above their ground states.

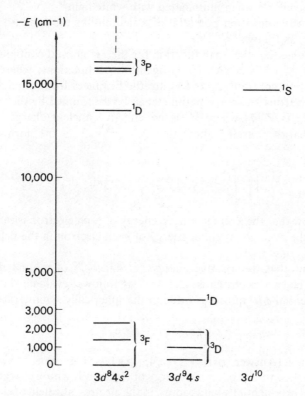

Fig. 5.4. Terms of low-lying configurations in Ni.

In conclusion, let it be said that the student will obtain much greater benefit from studying a table of ground configurations (see appendix D) than from reading a prolonged description of the periodic table.

Problems

5.1. Electric-dipole transitions between terms of the $1snl$ configurations in helium satisfy the selection rules $\Delta S = 0$, $\Delta L = \pm 1$. Indicate all the allowed transitions between the terms shown in Fig. 5.3. (In view of the rule $\Delta S = 0$ it is convenient to construct two separate energy-level diagrams, one for the triplets and one for the singlets and to draw in the transitions on each separately.) With the help of table 5.1 calculate the wavenumber (cm^{-1}) and wavelength (Å) of the transition $1s^2\ {}^1S$—

$1s2p$ ^1P. Which transitions give rise to the red line, 6,678 Å, to the yellow, 5,875 Å, and to the near ultraviolet line, 3,888 Å? In particular, which of these belong to the singlet, and which to the triplet spectrum? Which terms are metastable, and what are their excitation energies in eV? Will a bulb of helium gas at room temperature give rise to an absorption spectrum in the visible when illuminated with white light?

5.2. Find the ionization potential of a helium-like atom by a variation procedure, as follows:

(a) Assume that the wave function for the $1s^2$ ground configuration is a product, N $e^{-\alpha r_1}$ $e^{-\alpha r_2}$, of hydrogen-like $1s$ functions, where N is a normalization constant, r_1 and r_2 are the distances from the nucleus to the two electrons and α is a parameter to be determined by the variation method. (α is Z'/a_0 where Z' is an effective nuclear charge, not the nuclear charge Z itself.) Show that $N = \alpha^3/\pi$ from the normalization condition

$$(4\pi)^2 \int_0^\infty \int_0^\infty N^2 e^{-2\alpha r_1} e^{-2\alpha r_2} r_1^2 r_2^2 \, dr_1 \, dr_2 = 1.$$

(b) Show that the average kinetic energy of each electron is $(\hbar^2/2m)\alpha^2$, and that the average potential energy of each electron in the field of the nucleus is $-Ze^2\alpha/4\pi\varepsilon_0$.

(c) Show that the average energy $\langle e^2/4\pi\varepsilon_0 r_{12}\rangle$ of mutual repulsion between the two electrons is $\frac{5}{8}\alpha e^2/4\pi\varepsilon_0$ as follows: evaluate the electrostatic potential $\phi(r)$ at radius r due to the spherically symmetrical charge density

$$-e\,|\psi(r_2)|^2 = -eN\,e^{-2\alpha r_2}$$

of electron 2. (Answer: $\phi(r) = (-e/4\pi\varepsilon_0 r)(1 -(1 + \alpha r)\,e^{-2\alpha r})$.) Then find the average potential energy of electron 1, whose charge density also varies exponentially with radius, in the electrostatic field of electron 2.

(d) From (b) and (c) set up the average of the total energy $\langle \mathcal{H}\rangle$ of the two-electron atom and minimize it with respect to α by putting $d\langle \mathcal{H}\rangle/d\alpha = 0$. Hence evaluate $\alpha(=Z'/a_0)$ in terms of Z. Evaluate $\langle \mathcal{H}\rangle_{min}$. This is the variation procedure.

(e) $\langle \mathcal{H}\rangle_{min}$ is the energy of the $1s^2$ ground state of He relative to He^{++}. Subtract from it the energy of the ground state of He$^+$ to find the ionization potential of He. Express the result in Rydbergs and in eV, and compare with the experimental value of 1·81 Rydbergs. The extension of this treatment to Li$^+$, Be^{++}, etc., is straightforward.

5.3. Find the ionization potential of a helium-like atom by first-order perturbation theory, where the zeroth-order Hamiltonian is

$$\mathcal{H}_0 = -\frac{\hbar^2}{2m}\nabla_1^2 - \frac{\hbar^2}{2m}\nabla_2^2 - \frac{Ze^2}{4\pi\varepsilon_0 r_1} - \frac{Ze^2}{4\pi\varepsilon_0 r_2}$$

and the perturbation is $\mathscr{H}' = e^2/4\pi\varepsilon_0 r_{12}$. The zeroth-order wave function for the $1s^2$ ground state is

$$N\, e^{-Zr_1/a_0}\, e^{-Zr_2/a_0},$$

that is, it is a product of true hydrogenic $1s$ functions in which the nuclear charge Z appears, rather than the screened charge Z' of problem 5.2. The evaluation of $\langle e^2/r_{12}\rangle$ proceeds as in problem 5.2(c), with Z' replaced by Z. Compare the result with that found by the variation method in problem 5.2 and with the experimental result.

5.4. For the $3p^5$ configuration of Cl write down the determinantal product function for which M_S is a maximum and, subject to this condition, M_L is also a maximum. What term (S, L) does this correspond to? Verify that these values of S and L are the same as for the ground term of Al.

*5.5. Consider a helium-like ion in which the nuclear charge is Z. In this problem we estimate the energy of the excited state $|1s; nl\rangle$, for $l \neq 0$, by treating $e^2/4\pi\varepsilon_0 r_{12}$ as a perturbation in first order without exchange. The result which we are to evaluate is $E = E_n + J$ (eq. (5.46)) measured relative to the $1s$ ground state of the corresponding hydrogen-like ion, where E_n is a zeroth-order energy for the nl electron and J is the direct integral given by eq. (5.47).

Assume that the unperturbed wave function is a product of unscreened hydrogen-like wave functions

$$u_{1s}(1) = R_{1s}(Z, r_1)Y_0^0(\theta_1 \phi_1),$$

$$u_{nl}(2) = R_{nl}(Z, r_2)Y_l^m(\theta_2 \phi_2).$$

(a) Write down an expression for E_n.

To evaluate the perturbation energy $\Delta E = J$ from the integral

$$\frac{4\pi\varepsilon_0}{e^2} J = \int\int \frac{|u_{1s}|^2\, d\tau_1\, |u_{nl}|^2\, d\tau_2}{r_{12}}$$

we use formulae which occur later in the book. We expand $1/r_{12}$:

$$\frac{1}{r_{12}} = \sum_{k=0}^{\infty} \frac{r_<^k}{r_>^{k+1}} P_k(\cos\omega) \qquad\qquad \text{(eq. (7.10))}$$

where $r_<$ is the lesser of r_1 and r_2, $r_>$ is the greater and ω is the angle between $\mathbf{r}_1$ and $\mathbf{r}_2$. We separate the co-ordinates by eq. (9.50)

$$\frac{1}{r_{12}} = \sum_{k=0}^{\infty} \frac{r_<^k}{r_>^{k+1}} \frac{4\pi}{2k+1} \sum_{q=-k}^{k} (-1)^q Y_k^{-q}(\theta_1\phi_1) Y_k^q(\theta_2\phi_2).$$

We can now deal with the angular integrations in J:

(b) Show by means of a qualitative argument that since electron 1 has $l = 0$ there is only one term in the sum over k, namely $k = 0$.

(c) Show that the direct integral reduces to the radial integral

$$\frac{4\pi\varepsilon_0}{e^2} J = \int_0^\infty \int_0^\infty R_{1s}^2(Z, r_1) r_1^2 \, dr_1 \, \frac{1}{r_>} R_{nl}^2(Z, r_2) r_2^2 \, dr_2$$

which is a Slater integral, F^0 (see p. 115).

(d) The notation in (c) means that

$$\frac{4\pi\varepsilon_0}{e^2} J =$$

$$\int_0^\infty R_{1s}^2(Z, r_1) r_1^2 \, dr_1 \left(\int_0^{r_1} \frac{R_{nl}^2(Z, r_2) r_2^2 \, dr_2}{r_1} + \int_{r_1}^\infty \frac{R_{nl}^2(Z, r_2) r_2^2 \, dr_2}{r_2} \right).$$

Show that, for $l \neq 0$, to a certain approximation the first term ≈ 0 and the second term $\approx \langle r^{-1} \rangle_{nl}$.

(e) Hence, show that the final energy is approximately

$$E = -\frac{Z^2}{n^2} + \frac{2Z}{n^2} \quad \text{Rydbergs.}$$

(f) Comment on the adequacy of first-order perturbation theory in this calculation, particularly for $Z = 2$, and contrast the method with that using the effective potential of eq. (5.49).

(g) Calculate the direct integral (part (d)) with hydrogen-like functions for the level $1s2p$, and compare the result with the approximate answer of part (e), with the answer derived from eq. (5.49) and with the experimental result (table 5.1).

6. The central-field approximation

We have treated in some detail the electrostatic interactions for a two-electron atom. We now generalize the Schrödinger equation (5.1) for two electrons to describe the electrostatic interactions for N electrons:

$$\left\{ \sum_i^N \left(-\frac{\hbar^2}{2m} \nabla_i^2 - \frac{Ze^2}{4\pi\varepsilon_0 r_i} \right) + \sum_{i>j} \frac{e^2}{4\pi\varepsilon_0 r_{ij}} \right\} \psi = E\psi, \tag{6.1}$$

in which the mutual repulsion term is summed over all pairs of electrons. The presence of this two-electron operator prevents an immediate separation of the wave function into one-electron functions. Furthermore the mutual repulsion is in general too large to be treated as a perturbation, that is, the zeroth-order approximation in which $\sum_{i>j} e^2/4\pi\varepsilon_0 r_{ij}$ is neglected is not at all realistic (see Fig. 6.1a).

The procedure is to regroup the terms in the Hamiltonian of eq. (6.1) in a physically more significant way.

6.1. The central field

The strong effect of the attractive potential terms $\sum_i (-Ze^2/4\pi\varepsilon_0 r_i)$ is considerably reduced by the central part of the repulsive terms $\sum_{i>j} e^2/4\pi\varepsilon_0 r_{ij}$, that is by the part of $\sum_{i>j} e^2/4\pi\varepsilon_0 r_{ij}$ representing a force on an electron directed away from the central charge Ze. We have already seen in chapter 5 that this partial cancellation is manifested as a screening, from the outer electrons, of the central charge by the inner electrons. Let the central part of $\sum_{i>j} e^2/4\pi\varepsilon_0 r_{ij}$ be assumed to be $\sum_i S(r_i)$. Then the total central potential is $\sum_i (-Ze^2/4\pi\varepsilon_0 r_i + S(r_i))$ which we shall call $\sum_i U(r_i)$; $\sum_{i>j} e^2/4\pi\varepsilon_0 r_{ij} - \sum_i S(r_i)$ is left over. We now rewrite the Hamiltonian of eq. (6.1):

$$\mathcal{H} = \mathcal{H}_0 + \mathcal{H}_1, \tag{6.2}$$

where

$$\mathcal{H}_0 = \sum_i \left(-\frac{\hbar^2}{2m} \nabla_i^2 + U(r_i) \right), \tag{6.3}$$

and

$$\mathcal{H}_1 = \sum_{i>j} \frac{e^2}{4\pi\varepsilon_0 r_{ij}} - \sum_i \left(\frac{Ze^2}{4\pi\varepsilon_0 r_i} + U(r_i) \right). \tag{6.4}$$

Now we hope that $\mathcal{H}_1 \ll \mathcal{H}_0$, and that perturbation theory can be applied: the justification for this hope is to be found in the successful applications. The contrast between the use of (a) $V = \sum_i (-Ze^2/4\pi\varepsilon_0 r_i)$ and (b) $V = \sum_i U(r_i)$ as the zeroth-order potential is shown qualitatively in Fig. 6.1.

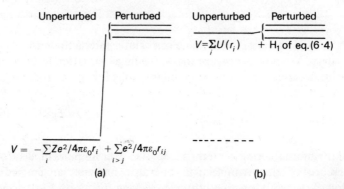

Fig. 6.1. Schematic energy level diagram, contrasting the use of (a) $-\sum_i Ze^2/4\pi\varepsilon_0 r_i$ and (b) $\sum_i U(r)$ as zeroth-order potentials.

In the zeroth approximation we neglect $\mathcal{H}_1$ in comparison with $\mathcal{H}_0$ (this is the central field approximation) and we have to solve

$$\mathcal{H}_0 \psi = \sum_i \left(-\frac{\hbar^2}{2m}\nabla_i^2 + U(r_i) \right) \psi = E\psi. \tag{6.5}$$

Since we have assumed that $\mathcal{H}_0$ is of the form $\sum_i \mathcal{H}_{0i}$, ψ is separable into a product of single-electron functions, or rather into the form of a Slater determinant (to satisfy the requirement of anti-symmetry), and the total energy is the sum of single-electron energy eigenvalues:

$$E = \sum_i E_{n_i l_i}. \tag{6.6}$$

Thus, for each electron

$$\left(-\frac{\hbar^2}{2m}\nabla^2 + U(r) \right) \psi_{nlm_l m_s} = E_{n,\,l}\,\psi_{nlm_l m_s}. \tag{6.7}$$

Equation (6.7) is the same as for hydrogen, except that the more general *central-field* potential $U(r)$ replaces $-Ze^2/4\pi\varepsilon_0 r$ and hence the energy depends on n and l; but the wave functions $\psi_{nlm_l m_s}$ are still degenerate with respect to m_l and m_s. In view of the central-field approximation the

6.1. The central field

$\psi_{nlm_lm_s}$, are separable into radial, angular, and spin parts:

$$\psi_{nlm_lm_s} = R_{nl}(r)Y_l^{m_l}(\theta, \phi)\chi(m_s). \tag{6.8}$$

Since the angular part is the same as for hydrogen the discussion of angular momenta given in chapters 2 and 4 applies equally well here.

In this approximation the total wave function describes a *configuration* of electrons, by which is meant that the good quantum numbers n_i and l_i are specified for each electron. It is clear that the l_i are good quantum numbers because classically the electrons are each moving independently round a centre of attraction and their orbital angular momenta are separately constants of the motion.

The chief assumptions so far are that $U(r)$ is spherically symmetrical and that eq. (6.7) in which $U(r)$ appears is the same for all the electrons of a many-electron atom. The problem that remains is to solve a radial equation of the form of eq. (2.37) in which $U(r)$ is not yet known. The method of attack is due to Hartree and is called the self-consistent field method. We shall only give an outline of this method here.

First a reasonable guess is made for the spherically symmetrical potential $U(r_k)$ for the kth electron. As we have assumed already, the kth electron is regarded as moving in a potential produced by the nuclear charge and by the charges of all the other electrons. The radial part of eq. (6.7) is solved numerically for ψ_k, with ψ_k specified by nlm_lm_s. This is repeated for all the electrons. Now an iterative procedure begins. The $\psi_i(i \neq k)$ are used to work out the charge distribution of all the electrons except the kth so as to provide an improved $U(r_k)$ from

$$U(r_k) = -Ze^2/4\pi\varepsilon_0 r_k + \sum_{i \neq k} \overline{\int \frac{e}{4\pi\varepsilon_0 r_{ki}} (e\psi_i^*\psi_i)\, d\tau} \tag{6.9}$$

in which the new $U(r_k)$ is forced to be spherically symmetrical by an averaging over all angles in the second term of eq. (6.9) as indicated by a bar. The whole procedure is repeated until the final results converge to self-consistent values of the $U(r_k)$ and ψ_k. The wave function for the whole atom is taken to be a simple product function in Hartree's method:

$$\psi = \prod_k \psi_k \tag{6.10}$$

which is *not* a properly anti-symmetrized Slater determinant. The exclusion principle is taken into account only in so far as the energy of the ground state is taken to be the lowest energy consistent with the assignment of different sets of quantum numbers nlm_lm_s to each electron.

This intuitive procedure turns out to be equivalent to the use of the variational principle to minimize the total energy. In all variational methods the energy of the lowest excited state would be found by minimizing the energy subject to the additional restriction that the excited-state wave function is to be orthogonal to the ground-state wave function.

The use of properly anti-symmetrized wave functions in the solution of the radial problem leads to the so-called Hartree–Fock method. This approximation is somewhat better in that electrostatic exchange terms are included in the treatment, but the calculations are certainly longer and more difficult.

It has been implicitly assumed that equivalent electrons, those having the same values of n and l, are described by the same radial function for all possible m_l and m_s and that only the angular parts of the wave function differ. Relaxing this restriction might lead to a still better approximation, and a little recent work has in fact been done in this kind of computation. For some applications it would be desirable to have relativistic Hartree–Fock functions, in which the Dirac equation rather than the Schrödinger equation is used in the single-electron problem. However, hardly any work has been done in this direction. At such levels of sophistication the calculations become very complex, and it is only with the use of high-speed computers that progress has been made.

Radial wave functions are used for calculating expectation values of functions of r, and in applying the wave functions one has to bear in mind the assumptions on the basis of which they are derived. Wave functions based on the self-consistent field method are designed to give the best values of the total electrostatic energy of an atom; small energy differences are of course given less accurately. Quantities which are more sensitive than total energy to the form of the wave function are not necessarily given at all well by approximate wave functions which give the best total energy. Wave functions which lead, furthermore, to a good fit with experiment for $\int \psi_a^* \mathbf{r} \psi_b \, d\tau$, the integral which is needed to calculate electric-dipole transition probabilities, may lead to inaccurate values of $\langle r^{-3} \rangle$, the quantity which occurs in the formulae for fine-structure and hyperfine-structure splitting in a central field. Clearly a calculated value of $\langle r^{-3} \rangle$ is very sensitive to the behaviour of the wave function at small values of r; on the other hand the region of small r contributes very little to the integral $\int \psi_a^* \mathbf{r} \psi_b \, d\tau$. It is unfortunate but true that one has to beware of relying on *ab initio* radial calculations of quantities other than energy for many-electron atoms to an accuracy better than 10 or 20 per cent. It has sometimes been found in the past that calculated total energies have been accurate (~ 1 per cent) but that transition probabilities have been wrong by large factors.

For some limited applications wave functions of an analytic form, such as modified hydrogenic functions with one or two adjustable parameters, have been used. These serve as approximations when more accurate wave functions are not known, and they have the advantage that expectation values of functions of r can be calculated in closed form. Such functions have been used particularly for the cases $l = n - 1$, for example $2p$ or $4f$, in which the radial wave function is a bell-shaped curve with no nodes.

6.2. Thomas–Fermi potential

In the last section we have spoken of making a reasonable guess at a spherically symmetrical potential to be used as a starting point for the self-consistent field method. One way of doing this is to make use of the Thomas–Fermi model.

This model pictures the electrons in an atom as a Fermi gas, that is a statistical assembly of electrons obeying Fermi–Dirac statistics, in which the exclusion principle is taken into account. The purpose of the model is to provide a method of calculating the electron density and from it the electrostatic potential due to the nucleus and the cloud of electrons.

The calculation proceeds as follows: we assume that electrons are moving in a box of volume dv in co-ordinate space which is large enough to contain many electrons but at the same time small enough that the electrostatic potential $\phi(r)$ does not vary appreciably over the size of the box. Under this assumption the electrons are moving freely, with no forces acting on them, and their translational momenta can be taken to be directed isotropically in momentum space. Thus the volume of momentum space available to electrons with absolute value of momentum $\leqslant p$ is $(4\pi/3)p^3$, the volume of a sphere of radius p; and the volume of phase space available is $(4\pi/3)p^3 \, dv$.

The exclusion principle states, in this context, that not more than two electrons are allowed in each volume of size h^3 in phase space. We now assume further that the electrons are packed in phase space as densely as possible consistent with the exclusion principle, that is, their kinetic energy is a minimum, which is equivalent to working at the absolute zero of temperature. Under these conditions the number of electrons per unit volume with momentum less than a maximum value p_0 is

$$n = \frac{2}{h^3} \times \frac{4\pi}{3} p_0^3, \tag{6.11}$$

or in terms of a maximum kinetic energy $T_0 = p_0^2/(2m)$

$$n = \frac{8\pi}{3h^3} (2mT_0)^{3/2}. \tag{6.12}$$

The electrostatic potential energy for an electron is $-e\phi$, and the condition that an electron does not escape from the atom is

$$T - e\phi \not> 0, \tag{6.13}$$

whence the maximum kinetic energy is given by

$$T_0 = e\phi. \tag{6.14}$$

101

The charge density $\rho = -en$ is therefore expressed in terms of the potential ϕ, from eqs. (6.12) and (6.14):

$$\rho = -\frac{8\pi}{3h^3} e(2me\phi)^{3/2}. \tag{6.15}$$

The charge density ρ is, like ϕ, a function of r. The two are related by Poisson's equation

$$\nabla^2\phi = -\rho/\varepsilon_0, \tag{6.16}$$

which becomes, with eq. (6.15), a differential equation in ϕ:

$$\nabla^2\phi = \frac{4}{3\pi h^3} \frac{e(2me\phi)^{3/2}}{4\pi\varepsilon_0}. \tag{6.17}$$

For this atomic model we require a solution of eq. (6.17) such that

$$\lim_{r\to 0} \phi(r) = Ze/4\pi\varepsilon_0 r \tag{6.18}$$

where Ze is the nuclear charge, and

$$\lim_{r\to\infty} r\phi(r) = 0 \tag{6.19}$$

which ensures that the atom as a whole is uncharged. Equation (6.17) is usually rewritten with the following changes:

$$\phi(r) = Z_{\text{eff}} e/4\pi\varepsilon_0 r = \chi(r)Ze/4\pi\varepsilon_0 r, \tag{6.20}$$

where

$$\chi(r) = Z_{\text{eff}}/Z, \tag{6.21}$$

and

$$r = bx, \tag{6.22}$$

where

$$b = (3\pi)^{2/3}\, 2^{-7/3} \frac{4\pi\varepsilon_0 h^2}{me^2} Z^{-1/3} \approx 0{\cdot}885a_0 Z^{-1/3}. \tag{6.23}$$

Thus eq. (6.17) becomes

$$\frac{\mathrm{d}^2\chi}{\mathrm{d}x^2} = x^{-1/2}\chi^{3/2}. \tag{6.24}$$

This is a universal equation which may be solved numerically once and for all to give χ as a function of x. Before we discuss a particular example we notice from eq. (6.20) that the potential $\phi(r)$ has been expressed as a screened Coulomb potential with Z_{eff} as an effective screened nuclear

6.2. Thomas–Fermi potential

charge. Also, from eqs. (6.22) and (6.23), we see that b is just a scaling factor for distance from the nucleus: whereas eq. (6.24) which is independent of Z indicates that the form of the potential, and hence of the electron charge distribution, is the same for all Thomas–Fermi atoms, the size of an atom actually decreases slowly as $Z^{-1/3}$.

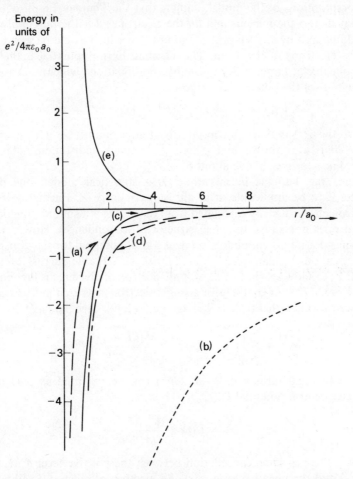

Fig. 6.2. Potential curves for comparison with the Thomas–Fermi potential at $Z = 20$. (a) $-e^2/4\pi\varepsilon_0 r$; (b) $-20e^2/4\pi\varepsilon_0 r$; (c) $V(r) = (-20e^2/4\pi\varepsilon_0 r)\,\chi\,(r;\,20)$; (d) $(-e^2/4\pi\varepsilon_0 r)\,\{19\,\chi\,(r;\,20)\,+\,1\}$; (e) $\hbar^2 6/2mr^2$.

In Fig. 6.2 is plotted the Thomas–Fermi potential energy $V(r) = -e\phi(r)$ in units of $e^2/4\pi\varepsilon_0 a_0$ as a function of r/a_0 for $Z = 20$. The data are taken from the table of χ versus x given in Condon and Shortley.† In the same figure $-e^2/4\pi\varepsilon_0 r$ ($Z_{\mathrm{eff}} = 1$) and $-20e^2/4\pi\varepsilon_0 r$ ($Z_{\mathrm{eff}} = 20$) are also

† E. U. Condon and G. H. Shortley, *The Theory of Atomic Spectra*, C.U.P. 1951, p. 337.

plotted. The effect of electrostatic screening of the nuclear charge is clearly shown for the Thomas–Fermi potential, which is intermediate between the two extremes ($Z_{eff} = 1$ and $Z_{eff} = 20$) in the region $r \sim a_0$. As $r \to 0$, $V(r) \to -Ze^2/4\pi\varepsilon_0 r$. But as $r \to \infty$, $V(r)$ approaches zero much more rapidly than $-e^2/4\pi\varepsilon_0 r$. This behaviour is implicit in one of the assumptions of the model, namely that the Thomas–Fermi potential is that due to the nucleus and *all* the electrons: it is the potential which would be seen by a small probe, and not that which would be seen by one of the electrons of the atom. The Thomas–Fermi potential is therefore inapplicable at larger r. A reasonable modification† which improves the behaviour of the potential at large r is

$$V(r) = (-e^2/4\pi\varepsilon_0 r)\{(Z - 1)\chi(r; Z) + 1\} \qquad (6.25)$$

where the $\chi(r)$ is that appropriate to atomic number Z. This curve for $Z = 20$ is also shown in Fig. 6.2: it becomes indistinguishable from $-e^2/4\pi\varepsilon_0 r$ for r/a_0 above about 6.

Since the Thomas–Fermi model is a statistical model, one might expect it to be applicable, if at all, only for large Z. Furthermore, the electron charge density calculated from the model is a smooth function of r and does not show any shell structure (l-dependence). However, the Thomas–Fermi potential does serve as a trial potential for self-consistent field methods. It can even be used, with remarkable success even for small Z, to show at what value of Z an electron of given n, l is first bound. If the radial part of eq. (6.7) for a single electron is written in its so-called reduced form, with $P(r) = rR(r)$ as in eq. (2.49), then we have

$$\frac{d^2 P(r)}{dr^2} + \frac{2m}{\hbar^2}\left(E - U(r) - \frac{\hbar^2}{2m}\frac{l(l + 1)}{r^2}\right) P(r) = 0. \qquad (6.26)$$

The centrifugal term $\hbar^2 l(l + 1)/(2mr^2)$ can be regarded as part of an effective central potential $U'(r)$:

$$U'(r) = U(r) + \frac{\hbar^2}{2m}\frac{l(l + 1)}{r^2}. \qquad (6.27)$$

Clearly there is some cancellation between the positive term $\hbar^2 l(l + 1)/(2mr^2)$ and the negative term $U(r)$. An electron of given n, l will not be bound unless $U'(r)$ is negative at some value of r. Bethe‡ indicates the way in which the use of the Thomas–Fermi potential for $U(r)$, the ground state potential, gives approximately the values $Z = 5, 21, 58, 124$ at which electrons with $l = 1, 2, 3, 4$ respectively are first bound in the periodic table. We have already mentioned in section 5.5 how a 3d electron is bound at $Z = 21$, but not at $Z = 20$. Figure 6.2 includes a plot of $\hbar^2 l(l + 1)/$

---

† A number of other improvements have been made to the original model.
‡ H. A. Bethe, *Intermediate Quantum Mechanics*, Benjamin.

$(2mr^2)$ against r/a_0 for $l = 2$. It is touch-and-go whether $U'(r) = V(r) + \hbar^2 6/2mr^2$ is positive or negative at $Z = 20$.

6.3. The gross structure of the alkalis

The alkali metals have electron configurations corresponding to closed shells plus one further electron. For example, the ground configuration of sodium ($Z = 11$) is $1s^2 2s^2 2p^6 3s$. Because of the spherical symmetry of the closed shells, the central field approximation is an extremely good description of the motion of the single valence electron in the field of the nucleus and of the other electrons. Since the core of closed shells contributes nothing to the angular momentum of the atom the energy levels can be labelled by the angular momentum l of the valence electron, together with its principal quantum number n. To describe the gross structure of the energy levels we omit spin-dependent interactions and consider only the central electrostatic field.

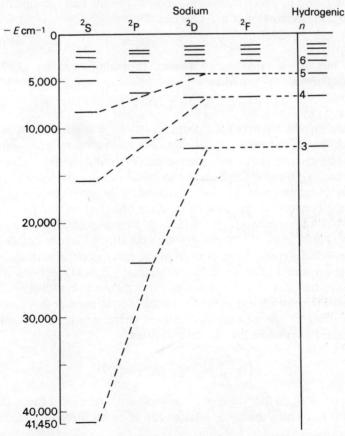

Fig. 6.3. Energy level diagram of Na. Terms with the same principal quantum number n are linked by dashed lines.

For large r the valence electron has a potential energy $-e^2/4\pi\varepsilon_0 r$: this expresses the fact that the nuclear charge Ze is screened by the core of $Z - 1$ electrons. For small r the potential energy approaches $-Ze^2/4\pi\varepsilon_0 r$ corresponding to an unscreened nucleus. Overall, the electrostatic attraction towards the nucleus is always greater than for hydrogen and so the energy levels are lower than the hydrogenic values for a given n. Moreover, the energy depends strongly on l. For small l the electron orbit is highly eccentric (to use the language of the old quantum theory): the electron penetrates the core and thus experiences a strong attraction towards the nucleus corresponding to a state of low energy. For larger l the penetration becomes less and the energy levels become more hydrogen-like. This is illustrated in Fig. 6.3 which shows an energy-level diagram for sodium.

One might attempt to express the energy levels in terms of an effective nuclear charge $Z_{eff}e$ to take account of the modification of the pure Coulomb field by penetration of the core. Z_{eff} would be a function of r for the valence electron, or of n, l. One could write

$$Z_{eff} = Z - \sigma(n, l) \tag{6.28}$$

where $\sigma(n, l)$ is a screening parameter. In analogy with eq. (2.57) for hydrogen the energy would be

$$E_{n, l} = -hcR \{Z - \sigma(n, l)\}^2/n^2, \tag{6.29}$$

an equation which is actually a definition of $\sigma(n, l)$ in terms of the experimental value of the energy $E_{n, l}$. It turns out that eq. (6.29) is more appropriate to a discussion of X-ray spectra than of optical spectra.

In studying X-ray spectra one is concerned with the energy required to remove from the atom an electron bound in an inner shell. The gross structure (neglecting spin effects) of these one-electron energy levels is treated in terms of a central field with strong electrostatic screening effects, that is, by eq. (6.29). $\sigma(n, l)$ is then the effective number of electron charges which screen the nucleus from the inner electron with quantum numbers n and l. For example, in sodium $(Z = 11)$ $E(1s) = -79\cdot4$ Rydbergs, hence $\sigma(1s) \approx 2\cdot1$. For a given n, l there is a smooth dependence of the one-electron X-ray energy levels on atomic number Z, practically independent of the periodicity which is characteristic of the valence electrons. Equation (6.29) may be re-written

$$\left(\frac{-E_{nl}}{hcR}\right)^{1/2} = \frac{1}{n}(Z - \sigma(n, l)). \tag{6.30}$$

A plot of $(-E_{nl}/hcR)^{1/2}$ against Z for various n, l is called a Bohr–Coster diagram,† in which the linear relationship of eq. (6.30) for a given n, l is

† For an account of X-ray spectra in some detail see H. G. Kuhn, *Atomic Spectra*, Section IV C.

fairly closely obeyed. The slope gives $1/n$, and the intercept on the Z-axis gives $\sigma(n, l)$. $\sigma(n, l)$ depends mainly on n and much less on l: hence the gross classification of X-ray shells by the quantum number n according to the names K, L, M . . . for $n = 1, 2, 3 \ldots$

To return to the problem of the single valence electron: historically an empirical formula which was found to fit the data for the alkalis was not eq. (6.29) but the Rydberg formula

$$E_{nl} = -hcR \frac{1}{(n - \delta(l))^2} \tag{6.31}$$

or

$$E_{nl} = -hcR/n^{*2} \tag{6.32}$$

where the effective principal quantum number n^* is $n - \delta$, and $\delta(l)$ is called the quantum defect. Equation (6.31) is obviously a modification of the Balmer formula for hydrogen. Unlike eq. (6.29) in which the effect of screening for an inner electron is written into Z_{eff}, eq. (6.31) has the form appropriate to an electron entirely outside the core ($Z_{eff} = 1$) with the effect of penetration of the core written into n^*. The virtue of the Rydberg formula in describing the excited states of a valence electron lies in the fact that $\delta(l)$ is very nearly *independent of n* for a given l. Thus for the ns configurations of sodium the quantum defects are

ns	$3s$	$4s$	$5s$	$6s$	$7s$
$\delta(s)$	1·374	1·357	1·353	1·351	1·350

For the np configurations $\delta(p) \approx 0·86$, for nd $\delta(d) \approx 0·01$, and for nf $\delta(f) \approx 0·00$. The large departure from zero of $\delta(s)$, and to a lesser extent of $\delta(p)$, indicates deep penetration of the core. On the other hand, the f-configurations are practically hydrogen-like. The relatively very small dependence of $\delta(l)$ on n may be attributed to higher order effects such as a polarization of the charge cloud of the core by the valence electron giving rise to a further attractive force towards the centre. The smaller the value of n, the closer the valence electron is to the core on the average and the greater is the degree of polarization. Thus on this account $\delta(l)$ is larger and E_{nl} even lower for smaller n.

The reader should study carefully tables of $\delta(l)$ or n^* for all the alkalis (such as tables 15(a) to 15(e) in Kuhn's *Atomic Spectra*) to see the variation of n^* with Z. For example the ground states of the alkalis, ns, show the following trend of n^*:

Element	Li	Na	K	Rb	Cs
n	2	3	4	5	6
n^*	1·588	1·626	1·771	1·805	1·869

Thus n^* changes very little compared with n, and the ionization potentials of these atoms are remarkably similar. In units of electron-volts they are:

Element	Li	Na	K	Rb	Cs
Ionization potential (eV)	5·40	5·14	4·34	4·17	3·89

For an iso-electronic sequence of alkali-like elements eq. (6.31) can be modified, again in analogy with hydrogen-like elements, to read

$$E_{nl} = -hcR \frac{Z_0^2}{\{n - \delta(l)\}^2} \qquad (6.33)$$

where $Z_0 e$ is equal to the charge of the nucleus plus that of the closed shell of electrons. $Z_0 = 1, 2, 3 \ldots$ for neutral, singly ionized, doubly ionized $\ldots$ atoms.

In such a sequence, $\delta(l)$ does depend on Z_0. For example, we consider the sequence iso-electronic with sodium for which $Z = 11$. The ionization potentials for this sequence, all of whose members have a $3s$ ground configuration, together with the quantum defect $\delta(s)$ defined by eq. (6.33) are given below:†

Element	Na I	Mg II	Al III	Si IV	P V	S VI	Cl VII
Ionization							
potential (eV)	5·14	15·03	28·44	45·13	65·01	88·03	114·27
$\delta(s)$	1·374	1·098	0·924	0·804	0·713	0·642	0·585

We see that $\delta(s)$ decreases along the sequence. This corresponds to the ions becoming more and more hydrogen-like. Such behaviour is to be expected because the perturbing effect of the closed shells of electrons becomes less and less important as the pure Coulomb attraction of the nuclear charge becomes more and more dominant.

The Rydberg formula, presented here as an empirical one, has in fact been reached as an approximate result of a self-consistent field treatment.

We shall return to a discussion of the fine structure in alkali spectra later.

Problems

6.1. Use the hydrogen radial wave functions of table 2.2 to evaluate the overlap integrals $\int_0^\infty R_{1s} R_{2p} r^2 \, dr$ and $\int_0^\infty R_{1s} R_{3p} r^2 \, dr$. Calculate also $\int_0^\infty R_{1s} r R_{2p} r^2 \, dr$ and $\int_0^\infty R_{1s} r R_{3p} r^2 \, dr$ and compare these off-diagonal elements with $\langle r \rangle_{1s}$, $\langle r \rangle_{2p}$, $\langle r \rangle_{3p}$ evaluated from table 2.3.

$$\left[\int_0^\infty x^n e^{-x} \, dx = n! \right]$$

† The notation Na I, Mg II, Al III, etc., is used to name the spectra of Na, Mg^+, Al^{++} etc. Na I is also called the arc spectrum of sodium, Mg II the first spark spectrum of magnesium and Al III the second spark spectrum of aluminium. This nomenclature is associated with the conditions in the light source: an arc is suitable for exciting neutral atoms and a spark discharge favours ionization and excitation of the resulting ions.

Problems

6.2. A radial wave function with no nodes ($l = n - 1$), corresponding to a Bohr circular orbit, might be written in analytical form as a modified hydrogenic function with one parameter:

$$R_{n,\, l \,=\, n\, -\, 1} = N r^{n-1} e^{-\alpha r}$$

subject to the normalization $\int_0^\infty R^2 r^2 \, dr = 1$.

By inspection, find the form of the parameter α which gives the energy in hydrogenic form, i.e., $E = -\frac{1}{2} Z'^2 e^2/(4\pi\varepsilon_0 n^2 a_0)$, for an electron moving in the field of an effective central charge $Z'e$.

Show that the peak of $R^2 r^2$, the probability density per unit radial thickness, lies at $r_0 = n^2 a_0/Z'$, which is an effective Bohr radius for the charge distribution.

Show also that with this wave function

$$\langle r^k \rangle = \frac{(2n + k)!}{(2n)! \, 2^k} (n a_0/Z')^k.$$

7. Angular problems in many-electron atoms

In the last chapter we discussed the central-field approximation for many-electron atoms, in which the zeroth-order Hamiltonian (eq. 6.3)) was

$$\mathcal{H}_0 = \sum_i \left(-\frac{\hbar^2}{2m} \nabla_i^2 + U(r_i) \right). \tag{7.1}$$

In that approximation the wave function for the atom was expressed as a linear combination of products of single-electron wave functions each of which was separable into radial, angular, and spin parts (eq. (6.8)). Different configurations, each specified by a set of quantum numbers $n_1 l_1, n_2 l_2, \ldots$, had different energies but there was degeneracy with respect to the magnetic quantum numbers m_{l_i} and m_{s_i}. In other words there were many states $|n_1 l_1 m_{l_1} m_{s_1}, n_2 l_2 m_{l_2} m_{s_2}, \ldots\rangle$ of a single configuration all of which had the same energy $E_{n_1 l_1, n_2 l_2}, \ldots$. In the central-field approximation we neglected the residual electrostatic term in the Hamiltonian which represents a non-central force (eq. (6.4)):

$$\mathcal{H}_1 = \sum_{i>j} \frac{e^2}{4\pi\varepsilon_0 r_{ij}} - \sum_i \left(\frac{Ze^2}{4\pi\varepsilon_0 r_i} + U(r_i) \right). \tag{7.2}$$

We now wish to consider the term $\mathcal{H}_1$ as a small perturbation together with other interaction terms which we have also neglected so far (see the list at the beginning of chapter 5). The largest of these other terms is the spin–orbit interaction

$$\mathcal{H}_2 = \sum_i \xi(r_i) \mathbf{l}_i \cdot \mathbf{s}_i. \tag{7.3}$$

Before we consider the interactions $\mathcal{H}_1$ and $\mathcal{H}_2$ in first-order perturbation theory let us dispose of second-order effects, not because they are unimportant in atomic spectra—far from it—but because we shall not be able to discuss them in detail within the scope of this book. In eqs. (7.2) and (7.3) we are dealing with two kinds of operators: a sum of one-electron operators, $F = \sum_i f_i$, like eq. (7.3) and the second term in eq. (7.2), and a sum of two-electron operators, $G = \sum_{i>j} g_{ij}$, like $\sum_{i>j} e^2/4\pi\varepsilon_0 r_{ij}$. In the representation appropriate to the central-field approximation, that is when the wave function for an N-electron atom is of the form of a determinantal product of N one-electron functions, these operators can have off-diagonal matrix elements. If $|A\rangle$ and $|B\rangle$ represent two determinantal product eigenfunctions of $\mathcal{H}_0$ (eq. (7.1)) with eigenvalues E_A and E_B, where the labels A and B each stand for a list of N individual sets of quantum numbers $n_1 l_1 m_{l_1} m_{s_1} \ldots n_N l_N m_{l_N} m_{s_N}$, then $\langle A|\, F\, |B\rangle$ can be non-vanishing if $|B\rangle$ differs from $|A\rangle$ by not more than one individual set of quantum numbers. Also $\langle A|\, G\, |B\rangle$ can be non-vanishing if $|B\rangle$ differs from $|A\rangle$ by not more than two individual sets.† The result is that, subject to certain limitations imposed by angular momentum considerations, the perturbations F and G can mix different configurations into the zeroth-order central-field configuration according to eq. (B.7) in appendix B. For example, under the action of the term $\sum_{i>j} e^2/4\pi\varepsilon_0 r_{ij}$ the unperturbed function $|A\rangle$ is modified to become ψ where

$$\psi = |A\rangle + \sum_B |B\rangle \frac{\langle B|\sum_{i>j} e^2/4\pi\varepsilon_0 r_{ij} |A\rangle}{E_A - E_B}, \qquad (7.4)$$

so that some of the character of the configurations B is mixed into the configuration A.

In what follows we shall assume that the energy denominators $E_A - E_B$ are so large that this configuration mixing is negligible. In other words we shall assume that we are able to treat each configuration as if it were isolated from all the others. We are therefore retaining the central-field approximation in zeroth order and we are about to consider the perturbations $\mathcal{H}_1$ and $\mathcal{H}_2$ of eqs. (7.2) and (7.3) in first order.

7.1. The *LS* coupling approximation

We are concerned now with the structure of a single configuration arising from the application of $\mathcal{H}_1$ and $\mathcal{H}_2$ as perturbations, that is we are concerned with the energy differences resulting from a lifting of the degeneracy within a single configuration and not with a shift of the energy of the configuration as a whole. For a given configuration n and l are fixed for each electron and the degeneracy is entirely with respect to the m_l and m_s. In first-order perturbation theory we take diagonal matrix elements of the operators $\mathcal{H}_1$ and $\mathcal{H}_2$. It is shown in more advanced texts that the diagonal matrix element of the interaction $\sum_j e^2/4\pi\varepsilon_0 r_{1j}$ between a valence electron 1 with quantum numbers $nlm_l m_s$ and electron j of a closed shell, where the summation is taken over all members of the closed shell, is independent of m_l and m_s. Thus this term does not lead to a splitting, only to a shift of the energy of the whole configuration. The same is true of the

† See E. U. Condon and G. H. Shortley, *The Theory of Atomic Spectra*, chapter 6.

sum over closed shells of interactions between pairs of electrons (a) when both members of the pair are in the same closed shell and (b) when they are in different closed shells. The single-electron terms, $-\sum_i (Ze^2/4\pi\varepsilon_0 r_i + U(r_i))$, of $\mathcal{H}_1$ also give only a shift of the energy of the configuration. All this is to say that the spherical symmetry of the closed shells leads to a great simplification which one might feel intuitively, namely that in considering the energy *splitting* within a configuration arising from electrostatic interactions one need only consider the interactions $\sum_{i>j} e^2/4\pi\varepsilon_0 r_{ij}$ between pairs of *valence* electrons.

As regards the terms of $\mathcal{H}_2 = \sum_i \xi(r_i)\mathbf{l}_i \cdot \mathbf{s}_i$, which are one-electron operators, it can easily be shown that the sum over a closed shell of this interaction is zero (see problem 7.4). Thus for the spin–orbit interaction also, only the valence electrons are involved. The reason for neglecting the other relativistic terms which we treated together with the spin–orbit interaction in hydrogen is that these terms are not spin-dependent and do not contribute to a splitting. They only shift the energy of the configuration as a whole, and like all the electrostatic effects which contribute only to a shift they are now to be ignored.

We are therefore left with the perturbation

$$\mathcal{H}_1 + \mathcal{H}_2 = \sum_{i>j} e^2/4\pi\varepsilon_0 r_{ij} + \sum_i \xi(r_i)\mathbf{l}_i \cdot \mathbf{s}_i \qquad (7.5)$$

where the summations are taken only over valence electrons. In carrying through degenerate perturbation theory we need to investigate the constants of the motion for we must use a representation for the zeroth-order wave functions in which the perturbation is diagonal. It is therefore useful to make a list of the angular momentum operators with which the perturbations commute.

As in the discussion of helium in chapter 5 we can see that $\sum_{i>j} e^2/4\pi\varepsilon_0 r_{ij}$ commutes with $\mathbf{L}^2$ and the components of $\mathbf{L}$, in particular L_z, where $\mathbf{L} = \sum_i \mathbf{l}_i$ because the interaction is internal to the orbital system and cannot change the orbital angular momentum of the system as a whole; and of course it commutes with the spin operators $\mathbf{S}^2$ and S_z where $\mathbf{S} = \sum_i \mathbf{s}_i$ is the total spin angular momentum (as already remarked, we need only consider summation over the valence electrons). We can also form the total angular momentum of the electrons

$$\mathbf{J} = \mathbf{L} + \mathbf{S} = \sum_i \mathbf{l}_i + \sum_i \mathbf{s}_i = \sum_i \mathbf{j}_i \qquad (7.6)$$

where $\mathbf{j}_i = \mathbf{l}_i + \mathbf{s}_i$, the total angular momentum of a single electron. Then $\sum_{i>j} e^2/4\pi\varepsilon_0 r_{ij}$ commutes with $\mathbf{J}^2$ and J_z, a result which follows algebraically from the fact that $\sum_{i>j} e^2/4\pi\varepsilon_0 r_{ij}$ commutes with the components of $\mathbf{L}$ and $\mathbf{S}$ separately; also $\sum_{i>j} e^2/4\pi\varepsilon_0 r_{ij}$ does not give rise to an external torque on the electrons as a whole, so the total electronic angular momentum is a constant of the motion. But $e^2/4\pi\varepsilon_0 r_{ij}$ does not commute with

7.1. The *LS* coupling approximation

l_{z_i}: this is an expression of the fact that there is a torque on an individual electron due to its electrostatic interaction with another electron. The individual orbital angular momenta can also change their *magnitudes* under the action of the electrostatic repulsion, but the effect of this on the energy is in second order via eq. (7.4), an effect which we are assuming to be negligible. That is, for a single configuration isolated from other configurations the n_i and l_i are fixed. In passing, it should be pointed out that when configuration mixing does occur $\sum_{i>j} e^2/4\pi\varepsilon_0 r_{ij}$ can only mix states with the same *L*, *S*, and *J* because of the commutation relations; moreover, since $\sum_{i>j} e^2/4\pi\varepsilon_0 r_{ij}$ is even under a parity operation, only configurations of the same parity, given by $(-1)^{\sum_i l_i}$ can be mixed by this interaction.

By contrast, the spin–orbit operator $\sum_i \xi(r_i)\mathbf{l}_i \cdot \mathbf{s}_i$ commutes with $\mathbf{j}_i^2$ and $j_{z_i} = l_{z_i} + s_{z_i}$ as we have seen in section 4.3, but not with l_{z_i} and s_{z_i} separately. It also commutes with $\mathbf{J}^2$ and J_z since the interaction is internal to the atom, but not with $\mathbf{L}^2$, L_z, $\mathbf{S}^2$, or S_z.

In seeking a representation in which to treat these perturbations we need labels associated with constants of the motion. In the configuration representation a state is completely specified by four quantum numbers $nlm_l m_s$ for each electron. For an *N*-electron system one needs $4N$ independent labels. Thus for two electrons some of the possible choices are

(a) $(nlm_l m_s)_1 (nlm_l m_s)_2$;
(b) $(nl)_1 (nl)_2 LSM_L M_S$;
(c) $(nl)_1 (nl)_2 LSJM_J$;
(d) $(nl)_1 (nl)_2 (jm_j)_1 (jm_j)_2$;
(e) $(nl)_1 (nl)_2 j_1 j_2 JM_J$.

For three or more electrons twelve or more quantum numbers are needed and one begins to run out of angular momentum labels. In such cases it is necessary to classify the states according to other symmetry principles. To avoid these difficulties we shall confine our discussion mainly to two-electron systems.

Matters are greatly simplified if we are able to decide on the basis of experiment whether one of the perturbations we are considering can be neglected in comparison with the other. Although all three cases— spin–orbit interaction large, small, and of the same order compared with the residual Coulomb interaction—actually occur, the most common situation especially in light elements is that in which the spin–orbit interaction can be neglected as a first approximation. That is to say, the fine structure for which the spin–orbit interaction would be responsible is much smaller than the splitting into *terms* produced by the residual Coulomb interaction $e^2/4\pi\varepsilon_0 r_{12}$. As evidence of this we can refer to table 5.1 for helium: the single-triplet term differences quoted there are of the order of 1,000 cm^{-1}, but the fine structure, as we shall see later, is at most 1 cm^{-1}. For calcium ($Z = 20$) the ^{3}P and ^{1}P terms of the configuration $4s4p$ are separated by about 8,000 cm^{-1} while the fine structure

113

of the ^3P term extends over 150 cm^{-1}. In germanium ($Z = 32$) the ground configuration is $4p^2$ and its three terms ^3P, ^1D, and ^1S cover a region of 16,000 cm^{-1} while the fine structure of the ^3P term spreads over 1,500 cm^{-1}. In such examples, while the spin–orbit interaction increases with atomic number it is nevertheless small compared with the residual Coulomb interaction.

The approximation in which the spin–orbit interaction is neglected in comparison with the residual Coulomb interaction is called the *LS coupling approximation*. For, as we have seen, L and S are good quantum numbers, representing constants of the motion, under the perturbation $e^2/4\pi\varepsilon_0 r_{12}$.

For a single configuration in LS coupling either of the representations (b) or (c) above would be a suitable zeroth-order representation for the perturbation $e^2/4\pi\varepsilon_0 r_{12}$. That is, with $(nl)_1$ and $(nl)_2$ fixed, L and S are specified together with either M_L and M_S or J and M_J. The first-order energy shift is, in representation (b),

$$\Delta E = \langle (nl)_1 (nl)_2 LSM_L M_S | \; e^2/4\pi\varepsilon_0 r_{12} \; |(nl)_2 LSM_L M_S\rangle. \qquad (7.7)$$

$\overset{(nl)_1}{}$

It can be shown that ΔE is independent of the values of M_L and M_S: this is equivalent to the physically reasonable argument that the observed energy under such an interaction cannot possibly depend on the choice of the orientation of a set of co-ordinate axes in the laboratory. Thus there is $(2L + 1)(2S + 1)$-fold degeneracy with respect to M_L and M_S. The set of $(2L + 1)(2S + 1)$ states labelled by L and S is called a *term*. The energy depends only on L and S. In the $LSJM_J$ representation the energy is independent of J and M_J, and there is a $\sum_{J=|L-S|}^{L+S} (2J + 1)$-fold degeneracy. $\sum_{J=|L-S|}^{L+S} (2J + 1)$ is exactly equal to $(2L + 1)(2S + 1)$ because the number of states of a given term must be conserved independent of the choice of representation. The state $|LSJM_J\rangle$ can be expressed as a linear combination of the states $|LSM_L M_S\rangle$ having the same L and S but various combinations of M_L and M_S in a manner quite analogous to eq. (4.54).

The reason why the electrostatic energy seems to depend on S, even though $e^2/4\pi\varepsilon_0 r_{12}$ does not operate in spin space, is that the exchange symmetry requirements must be satisfied. This is just the question which was discussed in chapter 5 in connection with the energy levels of helium.

Actually to work out an energy shift like eq. (7.7) involves a knowledge of the radial wave functions of the single electrons. We cannot go into the details of such calculations in this book, but we can indicate the procedure for getting from eq. (7.7) to radial integrals. The state $|(nl)_1 (nl)_2 LSM_L M_S\rangle$ is first expanded in terms of determinantal product states. We have already seen an example of this in eq. (5.59). In this way the exchange symmetry requirements are satisfied and the wave function is expressed

7.1. The LS coupling approximation

in terms of single-electron functions. The matrix element, eq. (7.7), is then a sum of terms of the forms

$$J = \int u_a^*(1)u_b^*(2)\,(e^2/4\pi\varepsilon_0 r_{12})u_a(1)u_b(2)\chi_{m_s}^*(1)\chi_{m'_s}^*(2)\chi_{m_s}(1)\chi_{m'_s}(2)\,\mathrm{d}\tau, \quad (7.8)$$

and

$$K = \int u_a^*(1)u_b^*(2)\,(e^2/4\pi\varepsilon_0 r_{12})u_b(1)u_a(2)\chi_{m_s}^*(1)\chi_{m'_s}^*(2)\chi_{m'_s}(1)\chi_{m_s}(2)\,\mathrm{d}\tau. \quad (7.9)$$

These are just the direct and exchange integrals similar to eqs. (5.47) and (5.48), except that the angular and spin parts are included. The subscripts a and b each stand for a set of quantum numbers nlm_l. Since $e^2/4\pi\varepsilon_0 r_{12}$ does not operate on the spins and the χ functions are normalized and orthogonal, the products $\chi^*\chi$ can be set equal to unity (eq. (4.9)) with the provision that in the exchange integral K, $m'_s = m_s$. Otherwise the exchange integral vanishes.

The operator $e^2/4\pi\varepsilon_0 r_{12}$ is a function of the angle ω between $\mathbf{r}_1$ and $\mathbf{r}_2$. In the central-field approximation it is convenient to expand the operator in a series of Legendre polynomials:

$$1/r_{12} = (r_1^2 + r_2^2 - 2r_1 r_2 \cos\omega)^{-1/2} = \sum_{k=0}^{\infty} \frac{r_<^k}{r_>^{k+1}} P_k(\cos\omega) \quad (7.10)$$

where $r_<$ is the lesser of r_1 and r_2, and $r_>$ is the greater. The wave functions u_{nlm_l} are expressed as products of radial and angular parts, $R_{nl}(r)Y_l^{m_l}(\theta, \phi)$, or more conveniently $(1/r)P_{nl}(r)Y_l^{m_l}(\theta, \phi)$ where the normalization for the radial part is $\int_0^\infty [P_{nl}(r)]^2\,\mathrm{d}r = 1$ as in eq. (2.55). With the operator in the form of eq. (7.10) the integration over angles in eqs. (7.8) and (7.9) is carried out: this can be done once and for all, and the results can be tabulated as a set of coefficients which are functions of l, m_l for each pair of electrons. What is left in the diagonal matrix element of $e^2/4\pi\varepsilon_0 r_{12}$ is a set of radial integrals (so-called Slater integrals) F^k, which are direct integrals, and G^k, which are exchange integrals.† These have the forms

$$F^k = \frac{e^2}{4\pi\varepsilon_0} \int_0^\infty \int_0^\infty \frac{r_<^k}{r_>^{k+1}} [P_{n_1 l_1}(r_1)P_{n_2 l_2}(r_2)]^2\,\mathrm{d}r_1\,\mathrm{d}r_2$$

and

$$G^k = \frac{e^2}{4\pi\varepsilon_0} \int_0^\infty \int_0^\infty \frac{r_<^k}{r_>^{k+1}} P_{n_1 l_1}(r_1)P_{n_2 l_2}(r_1)P_{n_1 l_1}(r_2)P_{n_2 l_2}(r_2)\,\mathrm{d}r_1\,\mathrm{d}r_2.$$

These are the quantities which have to be worked out with a knowledge of radial wave functions. Actually, the quantities met in the literature

† Condon and Shortley treat this subject in chapters 6 and 7.

115

have a different notation. They are $F_k = F^k/D_k$ and $G_k = G^k/D_k$ where D_k is merely a number, depending on l_1, l_2, incorporated into the definition of F_k and G_k to make the arithmetic easier.

The index k extends only over a few small integral numbers which satisfy two conditions. The first is a triangular condition with the orbital angular momenta: if the numbers l, l' and k are represented as lengths, then these three lengths must be capable of forming the sides of a triangle, or in other words $|l - l'| \leqslant k \leqslant l + l'$. The second condition is a perimeter rule: $k + l + l'$ is to be an even integer. These conditions are imposed by the requirement that the angular coefficients in the matrix element of $e^2/4\pi\varepsilon_0 r_{12}$ should not vanish. For the direct integral F_k the conditions are to be obeyed by the sets (k, l_1, l_1) and (k, l_2, l_2) so k itself is even in this case; and for the exchange integral G_k they are to be obeyed by the set (k, l_1, l_2). As k increases F_k and G_k tend to become smaller and smaller.

The term energies in LS coupling depend just on sums of F_k's and G_k's each with a coefficient which depends on single-electron orbital angular momentum quantum numbers. The simple example of the configuration $1snl$ in helium has the terms ^1L and ^3L, where $L = l$ and the term energies† are, as in eq. (5.46),

$$E(^1L) = F_0 + G_l, \tag{7.11}$$

$$E(^3L) = F_0 - G_l. \tag{7.12}$$

F_0 and G_l each depend on the radial wave functions of both the $1s$ and nl electrons.

The accuracy of *ab initio* radial calculations does not as a rule match the experimental precision with which the energy levels can be determined. Therefore the F_k's and G_k's are often treated as *parameters* to be determined from experiment. There are sometimes simple relations between energy differences which depend only on the angular coefficients and these can be used as a test of the validity of the LS coupling approximation without a knowledge of radial wave functions. We shall give an example of this in the next section.

7.2. Allowed terms in *LS* coupling

Having seen in outline how the radial integrals of the central-field approximation enter into the calculation of the term energies in LS coupling, let us leave this difficult topic and turn instead to a discussion of the labelling of the terms in LS coupling. This much simpler problem is more a matter of book-keeping.

In considering the allowed values of L and S for non-equivalent

---

† A list of term energies for simple configurations is given in Condon and Shortley, chapter 7.

7.2. Allowed terms in *LS* coupling

electrons we find simple *branching rules*. If we start from a closed shell and add valence electrons one at a time we have the following scheme: the term for the closed shell is ^1S. Adding an nl electron we have the term ^2L, that is $S = \frac{1}{2}$ and $L = l$. Adding a further electron $n'l'$ to form the configuration $nln'l'$ we have all possible terms arising from the vector additions $\mathbf{S'} = \mathbf{S} + \mathbf{s'}$ and $\mathbf{L'} = \mathbf{L} + \mathbf{l'}$. The spin part gives $S' = S \pm \frac{1}{2}$, or since S was $\frac{1}{2}$ we have $S' = 0, 1$, that is singlets and triplets. For example, consider the configuration $npn'd$. The single np electron has the term ^2P ($L = l = 1$). Addition of the $n'd$ electron ($l' = 2$) gives singlets and triplets each with $L' = L + l', \ldots, |L - l'|$, or in this case $L' = 3, 2, 1$. Therefore the allowed terms of the $npn'd$ configuration are $^1F, \,^1D, \,^1P,$ $^3F, \,^3D, \,^3P$. The term ^2P of the np configuration is called the *parent* of all these six terms. The parent is indicated in the following way: $np(^2P)n'd\,^3F$; however, in this example it is not necessary to name the parent since there is no ambiguity.

When a third electron is added ambiguity does arise. The two-electron singlet parents beget doublets, and the triplet parents beget doublets and quartets. For example, out of the many terms of the configuration $npn'dn''p$, the $npn'd\,^1F$ parent gives terms $npn'd(^1F)n''p\,^2G, \,^2F, \,^2D$. ^2F terms also come from the parents $^1D, \,^3F$, and ^3D. All these four ^2F terms have different energies, but the same L and S. Specifying the parents helps to classify the terms.

For a quantitative example of the energies of terms in *LS* coupling let us consider the excited configuration $3p4p$ of silicon. From the branching rules given above, we find that the allowed terms are $^1S, \,^1P, \,^1D; \,^3S, \,^3P, \,^3D$. This configuration lies about 50,000 cm^{-1} above the ground configuration. It is a good example of a pure configuration, that is one which is not mixed with other configurations as a result of electrostatic interaction.† (Such examples in excited configurations are rather hard to find, for at excitation energies of the order of 50,000 cm^{-1} there are usually several other overlapping configurations so that the energy denominator of eq. (7.4) is not small.)

In terms of the Slater integrals for the $npn'p$ configuration the energies of the terms are given by

$$E(^1S; \,^3S) = F_0 + 10F_2 \pm (G_0 + 10G_2), \qquad (7.13)$$

$$E(^1P; \,^3P) = F_0 - 5F_2 \mp (G_0 - 5G_2), \qquad (7.14)$$

$$E(^1D; \,^3D) = F_0 + F_2 \pm (G_0 + G_2). \qquad (7.15)$$

In these equations F_0, F_2, G_0, and G_2 are treated as four unknown parameters and their values are found from a least-squares fit to the six known

---

† For this information and for the data on silicon which follow I am very grateful to B. Warner and R. D. Cowan.

experimental energies (F_0 includes the energy of the configuration†
relative to the ground state, and all energies are given relative to the
ground state as origin). The experimental term energies are shown in
Fig. 7.1, and from them are obtained the values

$$F_0 = 48,992\cdot152 \text{ cm}^{-1}$$
$$F_2 = 151\cdot149 \text{ cm}^{-1}$$
$$G_0 = 978\cdot852 \text{ cm}^{-1}$$
$$G_2 - 13\cdot819 \text{ cm}^{-1}.$$

Ignoring the small quantity G_2 which is much less than G_0, we see that the
effect of the exchange integral G_0 is to separate the singlet from the triplet
for each L by $2G_0 \approx 2,000 \text{ cm}^{-1}$. However, the triplets do not always lie
below the corresponding singlets: the order alternates with increasing L.
This alternation is a general result for two electrons. There are also certain
relations depending on the angular coefficients only; these are obtained
by taking ratios of differences of term energies, thus eliminating the
radial parameters. For example, from eqs. (7.13), (7.14), and (7.15)

$$\frac{E(^1S) - E(^1D)}{E(^1D) - E(^3P)} = \frac{9F_2 + 9G_2}{6F_2 + 6G_2} = \frac{3}{2}. \tag{7.16}$$

The experimental values give

$$\frac{51,612 - 50,189}{50,189 - 49,125} = \frac{1,423}{1,064} = 1\cdot34.$$

Similarly

$$\frac{E(^3S) - E(^3D)}{E(^3D) - E(^1P)} = \frac{9F_2 - 9G_2}{6F_2 - 6G_2} = \frac{3}{2}; \tag{7.17}$$

and experimentally

$$\frac{49,400 - 48,160}{48,160 - 47,284} = \frac{1,240}{876} = 1\cdot42.$$

Of course the experimental energy differences are differences of large
numbers so the ratios are very sensitive to small displacements of the
terms from their theoretical positions. Perturbation by spin–orbit interaction, which we are neglecting in LS coupling, can give rise to these
displacements. The size of the fine structure splitting of the ^3P and ^3D
terms is indicated in Fig. 7.1.

We have considered the six terms ^1S, ^1P, ^1D; ^3S, ^3P, ^3D arising from the
configuration $npn'p$. Each term is $(2S + 1)(2L + 1)$-fold degenerate:

---

† This energy of the configuration (or centre of gravity of the terms) is subject to shifts of
the configuration as a whole which, as mentioned at the beginning of section 7.1, we wish to
ignore. Our discussion is concerned only with the splitting of the configuration into terms.

therefore there are $\sum_{\text{terms}} (2S + 1)(2L + 1) = 36$ states of this configuration. The number of states is also $(2s_1 + 1)(2l_1 + 1)(2s_2 + 1)$ $(2l_2 + 1)$ which for $l_1 = l_2 = 1$ is again 36 as expected.

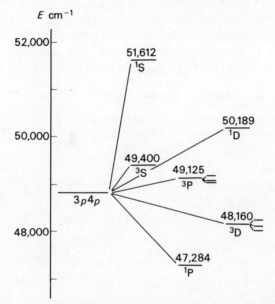

Fig. 7.1. Energies (above the ground state) of the terms of the excited configuration $3p4p$ in Si I. The fine-structure splitting of the triplet terms is also indicated.

We shall now consider the np^2 configuration as an example of two equivalent electrons. Because of the requirement of the exclusion principle that the two electrons which already have the same nl cannot have the same values of m_l, m_s, the number of states is greatly reduced. There are actually only 15 states of the np^2 configuration. The simplest way to find what terms are allowed is to adopt the following recipe:

(1) From the branching rules make a list of the *LS* terms allowed for the configuration of non-equivalent electrons, in this case $npn'p$: ^3D, ^3P, ^3S, ^1D, ^1P, ^1S.

(2) For each of these terms define the number N—$N(^3D)$, $N(^3P)$ etc.—which are allowed for the configuration of equivalent electrons, in this case np^2. These are the numbers which we wish to find.

(3) Write out the pairs of M_S, M_L values allowed for the *LS* terms under consideration (columns 1 and 2 of table 7.1). It is sufficient to list only non-negative values and it is convenient to start with maximum M_S and M_L and work downwards.

(4) Write out (column 3) all the determinantal product states allowed by the exclusion principle for which, in each row $\sum m_{s_i} = M_S$ and $\sum m_{l_i} = M_L$.

(5) Count the number of such states in each row and enter the result in column 4. This is the number of states in the $(m_l m_s)_1 (m_l m_s)_2$ representation having a given M_S and M_L. But this number must be *independent of representation*, so

(6) Write in column 5 the number $\sum_{L, S} N(^{2S+1}L)$ of states in the LS representation which are allowed to have the given M_S and M_L, and equate these to the numbers in column 4. This procedure gives enough simultaneous equations to evaluate the unknown values of N.

Table 7.1. Terms of the configuration np^2

M_S	M_L	Determinantal product states	Number $(m_l m_s)_1 (m_l m_s)_2$ Number $(LSM_L M_S)$
1	2	none	$0 = N(^3D)$
1	1	$\{\overset{+}{1}\ \overset{+}{0}\}$	$1 = N(^3D) + N(^3P)$
1	0	$\{\overset{+}{1}\ \overset{+}{-1}\}$	$1 = N(^3D) + N(^3P) + N(^3S)$
0	2	$\{\overset{+}{1}\ \overset{-}{1}\}$	$1 = N(^3D) + N(^1D)$
0	1	$\{\overset{+}{1}\ \overset{-}{0}\}, \{\overset{+}{0}\ \overset{-}{1}\}$	$2 = N(^3D) + N(^1D) + N(^3P) + N(^1P)$
0	0	$\{\overset{+}{1}\ \overset{-}{-1}\}, \{\overset{+}{-1}\ \overset{-}{1}\}, \{\overset{+}{0}\ \overset{-}{0}\}$	$3 = N(^3D) + N(^1D) + N(^3P) + N(^1P)$ $+ N(^3S) + N(^1S)$

In this simple case the answers can be obtained by inspection; they are $N(^3D) = 0$, $N(^3P) = 1$, $N(^3S) = 0$, $N(^1D) = 1$, $N(^1P) = 0$, $N(^1S) = 1$. Thus the only allowed terms of the configuration np^2 are ^1S, ^3P, and ^1D. In fact it can be shown generally that for two equivalent electrons nl^2 the quantity $S + L$ must be even. This is a useful mnemonic.

To find the particular linear combinations of the determinantal product states which give the terms in the $LSM_L M_S$ representation we need to know the properties of angular momentum operators.† Let $\mathbf{j}$ be a general angular momentum operator and let us consider a representation $|jm\rangle$ in which $\mathbf{j}^2$ and j_z are diagonal with eigenvalues $j(j + 1)$ and m respectively. The components j_x and j_y are best treated in the linear combinations

$$j_+ = j_x + ij_y, \tag{7.18}$$

$$j_- = j_x - ij_y, \tag{7.19}$$

for then

$$j_\pm |jm\rangle = \{j(j + 1) - m(m \pm 1)\}^{1/2} |jm \pm 1\rangle. \tag{7.20}$$

† See appendix C, and Dicke and Wittke chapter 9.

7.2. Allowed terms in LS coupling

The operators are called raising and lowering operators, or sometimes ladder or shift operators, because they transform the state $|jm\rangle$ into one in which the m-value is raised or lowered by unity. Further, we shall need to know how to apply these angular momentum operators to determinantal product functions. The operators we are dealing with here are single-particle operators f_i or sums $\sum_i f_i$ of such operators. It is a property of determinantal product functions that matrix elements, taken with these functions, of operators of the kind $\sum_i f_i$ are just the same as if simple product functions had been used. (For verification of this property see problem 7.4.)

The application of angular momentum operators to the transformation from the determinantal product representation to the LSM_LM_S representation is as follows (we take the np^2 configuration as an example): consider one of the unique identifications of table 7.1, say,

$$|^1D; 2, 0\rangle = \{\overset{+}{1}\,\overset{-}{1}\}, \tag{7.21}$$

where the notation on the left-hand side means that this state of the ^1D term has $M_L = 2$ and $M_S = 0$. Operating on eq. (7.21) with $L_- = l_{1-} + l_{2-}$ we obtain, on the left-hand side

$$L_-|^1D; 2, 0\rangle = (2(2+1) - 2(2-1))^{1/2}|^1D; 1, 0\rangle$$
$$= 2\,|^1D; 1, 0\rangle \tag{7.22}$$

and on the right-hand side

$$(l_{1-} + l_{2-})\{\overset{+}{1}\,\overset{-}{1}\} = (1(1+1) - 1(1-1))^{1/2}\{\overset{+}{0}\,\overset{-}{1}\}$$
$$+ (1(1+1) - 1(1-1))^{1/2}\{\overset{+}{1}\,\overset{-}{0}\}$$
$$= (2)^{1/2}\{\overset{+}{0}\,\overset{-}{1}\} + (2)^{1/2}\{\overset{+}{1}\,\overset{-}{0}\}. \tag{7.23}$$

(Equation (7.23) is an example of operating on a determinantal product function with a sum of single-electron operators.) Equating the results of eqs. (7.22) and (7.23) we have

$$|^1D; 1, 0\rangle = (2)^{-1/2}\{\overset{+}{0}\,\overset{-}{1}\} + (2)^{-1/2}\{\overset{+}{1}\,\overset{-}{0}\}. \tag{7.24}$$

With the operator L_- we have succeeded in 'laddering' down one rung in M_L from the state $|^1D; 2, 0\rangle$ to the state $|^1D; 1, 0\rangle$. The other state which was a linear combination of $\{\overset{+}{0}\,\overset{-}{1}\}$ and $\{\overset{+}{1}\,\overset{-}{0}\}$ in table 7.1 was $|^3P; 1, 0\rangle$. This must be orthogonal to $|^1D; 1, 0\rangle$, therefore

$$|^3P; 1, 0\rangle = \pm((2)^{-1/2}\{\overset{+}{0}\,\overset{-}{1}\} - (2)^{-1/2}\{\overset{+}{1}\,\overset{-}{0}\}). \tag{7.25}$$

121

There is an arbitrariness in the phase of this function, and in order to make our choice of phase consistent we can, alternatively, use a spin operator $S_+ = s_{1+} + s_{2+}$. For consider another unique identification

$$|^3P; 1, -1\rangle = \{\bar{1}\ \bar{0}\}. \tag{7.26}$$

$$S_+|^3P; 1, -1\rangle = (1(1+1) - (-1)(-1+1))^{1/2}|^3P; 1, 0\rangle$$
$$= (2)^{1/2}|^3P; 1, 0\rangle. \tag{7.27}$$

Also

$$(s_{1+} + s_{2+})\{\bar{1}\ \bar{0}\} = (\tfrac{1}{2}(\tfrac{3}{2}) - (-\tfrac{1}{2})(-\tfrac{1}{2}+1))^{1/2}\{\overset{+}{1}\ \bar{0}\}$$
$$+ (\tfrac{1}{2}(\tfrac{3}{2}) - (-\tfrac{1}{2})(-\tfrac{1}{2}+1))^{1/2}\{\bar{1}\ \overset{+}{0}\} = \{\overset{+}{1}\ \bar{0}\} + \{\bar{1}\ \overset{+}{0}\}. \tag{7.28}$$

Therefore, since $\{\bar{1}\ \overset{+}{0}\} = -\{\overset{+}{0}\ \bar{1}\}$ from the properties of a determinant, we have

$$|^3P; 1, 0\rangle = (2)^{-1/2}\{\overset{+}{1}\ \bar{0}\} - (2)^{-1/2}\{\overset{+}{0}\ \bar{1}\}. \tag{7.29}$$

We therefore choose the phase for the state $|^3P; 1, 0\rangle$ as that given by eq. (7.29). Indeed, there was an arbitrariness about the choice of phase in the fifteen original states of np^2: we could have attached a negative sign or a general factor $e^{i\delta}$ to any or all of them. But having made a consistent choice we must stick to it.

Further results are obtained by a continuation of the laddering process. For example, the results for the three states with $M_L = M_S = 0$ are

$$|^1D; 0, 0\rangle = (6)^{-1/2}\{\overset{+}{1}\ -\bar{1}\} + (6)^{-1/2}\{-\overset{+}{1}\ \bar{1}\} + 2(6)^{-1/2}\{\overset{+}{0}\ \bar{0}\}, \tag{7.30}$$

$$|^3P; 0, 0\rangle = (2)^{-1/2}\{\overset{+}{1}\ -\bar{1}\} - (2)^{-1/2}\{-\overset{+}{1}\ \bar{1}\}, \tag{7.31}$$

$$|^1S; 0, 0\rangle = (3)^{-1/2}\{\overset{+}{1}\ -\bar{1}\} + (3)^{-1/2}\{-\overset{+}{1}\ \bar{1}\} - (3)^{-1/2}\{\overset{+}{0}\ \bar{0}\}. \tag{7.32}$$

The $|^{2S+1}L; M_L, M_S\rangle$ states are properly normalized if the determinantal product states are.

Transformations like these are the ones required to express expectation values like eq. (7.7) in terms of integrals over single-electron functions like eqs. (7.8) and (7.9).

An example of the np^2 configuration is the ground configuration $3p^2$ of Si. In terms of the Slater integrals ($G_k = F_k$ for equivalent electrons) the energies of the terms are given by

$$E(^1S) = F_0 + 10F_2, \tag{7.33}$$

$$E(^3P) = F_0 - 5F_2, \tag{7.34}$$

$$E(^1D) = F_0 + F_2. \tag{7.35}$$

7.2. Allowed terms in *LS* coupling

A least-squares fit of F_0 and F_2 to the experimental data, which are shown in Fig. 7.2, gives

$$F_0 = 5{,}217{\cdot}322 \text{ cm}^{-1},$$
$$F_2 = 1{,}017{\cdot}420 \text{ cm}^{-1}.$$

The theoretical relation

$$\frac{E(^1S) - E(^1D)}{E(^1D) - E(^3P)} = \frac{9F_2}{6F_2} = \frac{3}{2} \tag{7.36}$$

is to be compared with the experimental ratio

$$\frac{15{,}394 - 6{,}299}{6{,}299 - 150} = \frac{9{,}095}{6{,}149} = 1{\cdot}48.$$

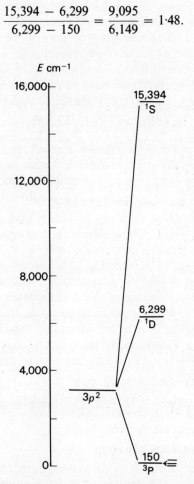

Fig. 7.2. Energies (above the ground state) of the terms of the ground configuration $3p^2$ in Si I. The fine-structure splitting of the triplet term is also indicated.

The agreement with theory is much better than in the case of $3p4p$ treated above. Comparison of Figs. 7.1 and 7.2 shows that the fine structure of the $3p^2$ ^{3}P term is much smaller relative to the electrostatic splitting than is the case in the $3p4p$ configuration. Hence one expects qualitatively that the LS coupling approximation is better for $3p^2$.

The ordering of the term energies in the $3p^2$ configuration is an example of *Hund's rule* which applies to the ground configuration of equivalent electrons in LS coupling. The rule is: The ground term has the largest value of S (maximum multiplicity $2S + 1$) consistent with the exclusion principle; if there are several such terms then the one with the largest value of L lies lowest in energy.

Thus for $3p^2$, of the three terms ^{1}S, ^{1}D, ^{3}P, the triplet lies lowest. Since there is only one triplet there is no need to invoke the rule about L. Actually, among the singlets the term ^{1}D with larger L lies below ^{1}S, but Hund's rule does not prescribe this ordering nor is it generally followed. For example, in Zr I the ground configuration is $4d^2 5s^2$ (the $5s^2$ electrons form an outer closed shell). The allowed terms of the $4d^2$ configuration are ^{1}S, ^{1}D, ^{1}G, ^{3}P, ^{3}F. Hund's rule states that the ground term is a triplet (larger S), and of the two triplets ^{3}F (larger L) is to be chosen. This is in fact the case, but among the singlets ^{1}D lies below ^{1}G.

On adding a third np-electron to the parent np^2 we come to a half-filled shell of equivalent p-electrons. We can discover the Hund's rule ground term by setting up a determinantal product state in which $M_S = \sum_i m_{s_i}$ is kept as large as possible consistent with the Pauli principle, at the same time trying to make $M_L = \sum_i m_{l_i}$ as large as possible. Thus for np^3 we have $\{\overset{+}{1}\,\overset{+}{0}\,\overset{+}{-1}\}$ for which $M_S = \frac{3}{2}$, $M_L = 0$. This is uniquely associated with the state $|^4\text{S}; 0, \frac{3}{2}\rangle$, so the ground term is ^{4}S. Examples of this configuration occur regularly through the periodic table: N, P, As, Sb, Bi. Similarly the ground term of nd^5 is ^{6}S, and of nf^7, ^{8}S. The ground term is bound to be an S-term ($M_{L_{max}} = 0$) if M_S is maximized. A further consequence of the special symmetry of half-filled shells is discussed in the next section.

The configuration nl^x of equivalent electrons for which the shell is more than half full (that is, $x > N/2$ where N is the maximum number of electrons, $2(2l + 1)$, allowed in the shell) has the same terms as the configuration nl^{N-x}. The Hund's rule ground term of the configuration np^4, for example, is associated with the state $\{\overset{+}{1}\,\overset{+}{0}\,\overset{+}{-1}\,\overset{-}{1}\}$ for which $M_S = 1$, $M_L = 1$. The ground term is ^{3}P, just as for two electrons: $\{\overset{+}{1}\,\overset{+}{0}\}$, $M_S = 1$, $M_L = 1$.

7.3. Fine structure in LS coupling

So far in this chapter we have considered the effect of the residual Coulomb interaction, eq. (7.2), in splitting a single configuration into terms. The good quantum numbers, representing constants of the motion, have been

7.3. Fine structure in *LS* coupling

L and S, and either M_L, M_S or J, M_J. We have chosen to work in the LSM_LM_S representation simply for convenience—for example, it has been easy to apply the shift operators $L_{\pm}$ and $S_{\pm}$. There has been no interaction between spin and orbit.

We now consider the spin–orbit interaction (eq. (7.3.))

$$\mathscr{H}_2 = \sum_i \xi(r_i)\mathbf{l}_i \cdot \mathbf{s}_i, \qquad \xi(r) = \frac{\hbar^2}{2m^2c^2} \cdot \frac{1}{r} \cdot \frac{dV}{dr} \qquad (7.37)$$

as a small perturbation. $\mathscr{H}_2$ can have matrix elements off-diagonal in L and S, leading to a breakdown of the LS coupling approximation, but we shall assume that the splitting due to spin–orbit interaction is very much smaller than the separation between terms so that we can consider a single term as an isolated system. Thus we retain the LS coupling approximation and treat L and S as good quantum numbers. We treat $\mathscr{H}_2$ by first-order perturbation theory in which we consider only matrix elements diagonal in L and S.

Because the zeroth-order state $|\gamma LSM_LM_S\rangle$, where γ specifies the configuration, is degenerate in M_L and M_S we need a representation in which the perturbation is diagonal. The arguments of section 4.3 for the single-electron case apply directly to the many-electron case in LS coupling. Just as in eqs. (4.54) and (4.55) we require the $\gamma LSJM_J$ representation; then the first-order energy shift is

$$\Delta E = \langle \gamma LSJM_J| \sum_i \xi(r_i)\mathbf{l}_i \cdot \mathbf{s}_i |\gamma LSJM_J\rangle. \qquad (7.38)$$

In eq. (7.38) we have to deal with an operator which is a sum of single-electron operators but whose diagonal matrix element is to be taken between states in a coupled representation. Let us first see what we can say about this with the help of the vector model.

For two electrons in LS coupling $\mathbf{l}_1$ and $\mathbf{l}_2$ form a vector resultant $\mathbf{L}$ about which they are considered (in a classical sense) to be precessing rapidly; similarly $\mathbf{s}_1$ and $\mathbf{s}_2$ precess rapidly about $\mathbf{S}$. The precession of $\mathbf{L}$ and $\mathbf{S}$ about their resultant $\mathbf{J}$ is much slower, corresponding to the assumption that the spin–orbit energy splitting is much smaller than the splitting due to residual electrostatic interaction (with exchange effects). That is to say, the classical motion of the orbital system $\mathbf{L} = \mathbf{l}_1 + \mathbf{l}_2$ is nearly independent of that of the spin system $\mathbf{S} = \mathbf{s}_1 + \mathbf{s}_2$, and $\mathbf{L}$ and $\mathbf{S}$ represent constants of the motion in this approximation (see Fig. 7.3). The classical vector model, set up in this dynamical way, is concerned with time averages: the component of $\mathbf{l}_1$ perpendicular to $\mathbf{L}$ averages to zero over the many cycles of the rapid precession of $\mathbf{l}_1$ about $\mathbf{L}$ which take place during one cycle of the slow precession of $\mathbf{L}$ about $\mathbf{J}$. Thus only the component of $\mathbf{l}_1$ lying along $\mathbf{L}$ is taken into consideration; similarly for $\mathbf{s}_1$.

The time average of $l_1 \cdot s_1$, indicated by a bar, with respect to the rapid precessional motion is

$$\overline{l_1 \cdot s_1} = \overline{\left(\frac{(l_1 \cdot L)}{L^2} L\right) \cdot \left(\frac{(s_1 \cdot S)}{S^2} S\right)}. \tag{7.39}$$

But $L \cdot S$ is constant under this averaging process, hence

$$\sum_i \xi_i \overline{l_i \cdot s_i} = \zeta(L, S)L \cdot S \tag{7.40}$$

where

$$\zeta(L, S) = \sum_i \xi_i \overline{\frac{(l_i \cdot L)(s_i \cdot S)}{L^2 S^2}}. \tag{7.41}$$

The rule is therefore: consider the most rapid precessional motion first, and project each of the individual vectors partaking in that motion on to their resultant. In transferring to quantum mechanics, the quantities L^2 and S^2 in eq. (7.41) are replaced by expectation values $L(L + 1)$ and $S(S + 1)$.

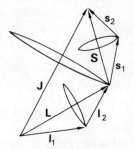

Fig. 7.3. Vector model for the angular momenta of two electrons in LS coupling.

This rule is a good one because quantum mechanics makes an equivalent statement in terms of matrix elements rather than time averages. To illustrate the projection procedure we refer to operators in orbit space. In taking matrix elements *diagonal in L* the dependence on M_L of *any vector operator in orbit space*, in particular l_i, is just that of the operator L itself. The reason for this is that all vector operators in orbit space transform in the same way under a rotation of the co-ordinate system so that the dependence on the orientation of the co-ordinate system (the M_L dependence) is the same for all. Thus for matrix elements diagonal in L we can write the direct proportionality:

$$\langle LM_L| \, l_i \, |LM_L'\rangle = c\langle LM_L| \, L \, |LM_L'\rangle \tag{7.42}$$

where the constant of proportionality c is independent of M_L and M_L'.

7.3. Fine structure in *LS* coupling

If it is understood that we are only to consider matrix elements diagonal in L, we can use the effective operator

$$\mathbf{l}_i = c\mathbf{L}. \tag{7.43}$$

The constant c can be evaluated in terms of a matrix element of $\mathbf{l}_i \cdot \mathbf{L}$ (for this evaluation we are at liberty to choose $M'_L = M_L$):

$$\langle LM_L| \mathbf{l}_i \cdot \mathbf{L} |LM_L\rangle = c\langle LM_L| \mathbf{L} \cdot \mathbf{L} |LM_L\rangle = cL(L + 1), \tag{7.44}$$

hence

$$c = \frac{\langle LM_L| \mathbf{l}_i \cdot \mathbf{L} |LM_L\rangle}{L(L + 1)}. \tag{7.45}$$

That this is independent of M_L can be verified for two electrons as follows: let $\mathbf{l}_i$ be $\mathbf{l}_1$, with $\mathbf{L} = \mathbf{l}_1 + \mathbf{l}_2$. Then

$$\mathbf{l}_2^2 = (\mathbf{L} - \mathbf{l}_1)^2 = \mathbf{L}^2 - 2\mathbf{l}_1 \cdot \mathbf{L} + \mathbf{l}_1^2 \tag{7.46}$$

or

$$\mathbf{l}_1 \cdot \mathbf{L} = \tfrac{1}{2}(\mathbf{L}^2 + \mathbf{l}_1^2 - \mathbf{l}_2^2). \tag{7.47}$$

Hence

$$\langle LM_L| \mathbf{l}_1 \cdot \mathbf{L} |LM_L\rangle = \tfrac{1}{2}\{L(L + 1) + l_1(l_1 + 1) - l_2(l_2 + 1)\}, \tag{7.48}$$

independent of M_L. In any case, it can be argued that the matrix element of a scalar operator within the orbit space cannot depend on the choice of orientation of a co-ordinate system.

What we have shown, therefore, is that eq. (7.42) with eq. (7.45) can be written†

$$\langle LM_L| \mathbf{l}_i |LM'_L\rangle = \frac{\langle LM_L| \mathbf{l}_i \cdot \mathbf{L} |LM_L\rangle}{L(L + 1)} \langle LM_L| \mathbf{L} |LM'_L\rangle \tag{7.49}$$

or, in words, the matrix element of $\mathbf{l}_i$ diagonal in L is proportional to the matrix element of $\mathbf{L}$ itself and the proportionality constant involves the expectation value of the projection of $\mathbf{l}_i$ upon $\mathbf{L}$. This is just the same as the result of the vector model. Thus finally the operator in eq. (7.38) can be replaced by the operator $\zeta(L, S)\mathbf{L} \cdot \mathbf{S}$ in the manner of the classical eq. (7.40) because we are only interested in its expectation value in the $LSJM_J$ representation—the matrix element of eq. (7.38) is diagonal in S and L. $\zeta(L, S)$ can be regarded as a parameter associated with the term ^{2S+1}L. Since

$$\mathbf{J}^2 = (\mathbf{L} + \mathbf{S})^2 = \mathbf{L}^2 + \mathbf{S}^2 + 2\mathbf{S} \cdot \mathbf{L} \tag{7.50}$$

† This equation states a special case of the more general Wigner–Eckart Theorem. All the M_L-dependence is in the matrix element $\langle LM_L|\mathbf{L}|LM'_L\rangle$.

or

$$\mathbf{S} \cdot \mathbf{L} = \tfrac{1}{2}(\mathbf{J}^2 - \mathbf{L}^2 - \mathbf{S}^2) \qquad (7.51)$$

we have

$$\Delta E = \langle \gamma LSJM_J | \zeta(L, S)\mathbf{L} \cdot \mathbf{S}|\gamma LSJM_J\rangle$$
$$= \tfrac{1}{2}\zeta(L, S)\{J(J + 1) - L(L + 1) - S(S + 1)\}. \qquad (7.52)$$

This result depends on J, and the degeneracy with respect to J has been lifted. Each fine structure component (L, S, J) of a term (L, S) is called a *level*. The energy of each level is still independent of M_J so there is still a $(2J + 1)$-fold degeneracy of each level with respect to M_J. Singlet terms are not split by the spin–orbit interaction (hence the name) for when $S = 0$, $J = L$ and in eq. (7.52) $\Delta E = 0$. Similarly S-terms are not split, because $L = 0$ and $J = S$.

Let us consider as an example a $^3\mathrm{P}$ term, for which $S = 1$, $L = 1$ and the possible values of J are 2, 1, 0. The energies of the levels $^3\mathrm{P}_2$, $^3\mathrm{P}_1$ and $^3\mathrm{P}_0$ are displaced from the $^3\mathrm{P}$ term energy by

$$\Delta E(^3\mathrm{P}_2) = \zeta(^3\mathrm{P}),$$
$$\Delta E(^3\mathrm{P}_1) = -\zeta(^3\mathrm{P}),$$
$$\Delta E(^3\mathrm{P}_0) = -2\zeta(^3\mathrm{P}),$$

from eq. (7.52). This is illustrated in Fig. 7.4, in which $\zeta(^3\mathrm{P})$ has been assumed to be positive. When ζ is positive the level with lowest J lies lowest and the fine structure *multiplet* is said to be *normal*. If ζ is negative the multiplet is *inverted*. We see that the energy intervals between $J = 2$ and 1 and between $J = 1$ and 0 are in the ratio of $2:1$.

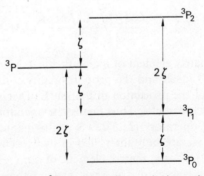

Fig. 7.4. The splitting of the $^3\mathrm{P}$ term according to the interval rule in LS coupling.

In general the energy difference between adjacent levels is, from eq. (7.52)

$$\Delta E(J) - \Delta E(J - 1) = \tfrac{1}{2}\zeta\{J(J + 1) - (J - 1)J\} = \zeta J. \qquad (7.53)$$

7.3. Fine structure in *LS* coupling

This equation expresses the *interval rule* in *LS* coupling: the energy difference between adjacent levels is proportional to the larger value of *J*. Departures from the interval rule indicate the presence of interactions other than the spin–orbit interaction, or more commonly a breakdown of the *LS* coupling approximation because the spin–orbit interaction cannot be considered as small compared with the residual electrostatic interaction. The spin–orbit interaction can have matrix elements off-diagonal in *L* and *S*, but diagonal in *J*, so it can mix together states of different *L* and *S* but of the same *J*. The result in second-order perturbation theory, according to eq. (B.5) of appendix B, is an energy shift

$$
\Delta E_2(SLJ)
$$
$$
= \sum_{S'L'}{}' \frac{\langle SLJ| \sum_i \xi_i \mathbf{l}_i \cdot \mathbf{s}_i |S'L'J\rangle \langle S'L'J| \sum_i \xi_i \mathbf{l}_i \cdot \mathbf{s}_i |SLJ\rangle}{E_{SL} - E_{S'L'}} \tag{7.54}
$$

which has the effect of shifting a level downward if it lies below the perturbing term in energy and upward if it lies above. In general, levels of the same *J* in different terms of a configuration *repel* each other in energy as a result of second-order perturbation. The term values are affected by this (see the experimental results following eq. (7.16)) and also the interval rule is violated.

To return to the example of the $3p^2$ ^{3}P ground term of Si I (Fig. 7.2): the fine-structure splitting parameter is $\zeta(^3P) = 74.02$ cm^{-1} and the energy level differences are $E(^3P_2) - E(^3P_1) = 146.16$ cm^{-1}, $E(^3P_1) - E(^3P_0) = 77.15$ cm^{-1}. Thus the interval rule is quite well obeyed. By contrast, in the term $3p4p$ ^{3}P (Fig. 7.1) $\zeta(^3P)$ would be 52.72 cm^{-1} in perfect *LS* coupling, but the energy differences are $E(^3P_2) - E(^3P_1) = 128.06$ cm^{-1} and $E(^3P_1) - E(^3P_0) = 32.38$ cm^{-1}. There is considerable departure from the interval rule.

For a single electron the quantity

$$
\xi(r) = \frac{\hbar^2}{2m^2c^2} \cdot \frac{1}{r} \cdot \frac{\mathrm{d}V}{\mathrm{d}r}
$$

of eq. (7.37) depends on the central potential *V*. In the nlm_lm_s representation the spin–orbit operator for one electron has the expectation value

$$
\langle nlm_lm_s| \xi(r)\mathbf{l} \cdot \mathbf{s} |nlm_lm_s\rangle = \zeta(nl)m_lm_s, \tag{7.55}
$$

where

$$
\zeta(nl) = \int_0^\infty R_{nl}\xi(r)R_{nl}r^2 \, \mathrm{d}r. \tag{7.56}
$$

$\zeta(nl)$ is a single-electron parameter depending on the radial part of the wave function. For a pure Coulomb field $V(r) \propto 1/r$ and $1/r \cdot \mathrm{d}V/\mathrm{d}r \propto$

$1/r^3$, so $\zeta(nl) \propto \langle r^{-3} \rangle$. The angular part of the integration in eq. (7.55) follows from the fact that the operator $\mathbf{l} \cdot \mathbf{s}$ can be expressed as

$$\mathbf{l} \cdot \mathbf{s} = l_z s_z + \tfrac{1}{2}(l_+ s_- + l_- s_+) \tag{7.57}$$

and of the three terms in eq. (7.57) only $l_z s_z$ has matrix elements diagonal in m_l and m_s.

Our next problem is to relate the $\zeta(SL)$ for a term to the $\zeta(n_i l_i)$ of the individual electrons of a configuration to which the term belongs. To do this we make use of a sum rule which depends on the invariance of the sum of diagonal matrix elements under a transformation from the $LSM_L M_S$ scheme to the $n_i l_i m_{l_i} m_{s_i}$ scheme. The rule states that for a given $M_L = \sum_i m_{l_i}$ and $M_S = \sum_i m_{s_i}$ the sum over terms of the diagonal matrix elements of $\mathcal{H}_2 = \sum_i \xi(r_i)\mathbf{l}_i \cdot \mathbf{s}_i$ in the $LSM_L M_S$ scheme is equal to the sum over determinantal states of the matrix elements of $\mathcal{H}_2$ in the $n_i l_i m_{l_i} m_{s_i}$ scheme. Now in the $\gamma LSM_L M_S$ representation

$$\langle \gamma LSM_L M_S | \mathcal{H}_2 | \gamma LSM_L M_S \rangle = \langle \gamma LSM_L M_S | \zeta(LS) \mathbf{L} \cdot \mathbf{S} | \gamma LSM_L M_S \rangle$$
$$= \zeta(LS) M_L M_S. \tag{7.58}$$

In the $n_i l_i m_{l_i} m_{s_i}$ scheme, since $\mathcal{H}_2$ is of the form of a sum of single-electron operators $\sum_i f_i$ whose properties we have discussed in connection with eq. (7.23), that is since

$$\mathcal{H}_2 = \sum_i \mathcal{H}_{2_i}, \tag{7.59}$$

the diagonal matrix element of $\mathcal{H}_2$ for an N-electron anti-symmetric function $|\psi_N\rangle$ is

$$\langle \psi_N | \mathcal{H}_2 | \psi_N \rangle = \langle \psi_N | \sum_i \mathcal{H}_{2_i} | \psi_N \rangle = \sum_i \langle n_i l_i m_{l_i} m_{s_i} | \mathcal{H}_{2_i} | n_i l_i m_{l_i} m_{s_i} \rangle$$
$$= \sum_i \zeta(n_i l_i) m_{l_i} m_{s_i}, \tag{7.60}$$

where eq. (7.55) has been used to evaluate the single-electron matrix elements. The sum rule relates eqs. (7.58) and (7.60) as follows:

$$\sum_{\text{terms}} \zeta(\gamma LS) M_L M_S = \sum_{\substack{\text{det.} \\ \text{states}}} \left(\sum_i \zeta(n_i l_i) m_{l_i} m_{s_i} \right); \quad M_L, M_S \text{ fixed.} \tag{7.61}$$

The application of eq. (7.61) is most easily seen when we have the terms and the determinantal states written out, as in table 7.1 for the configuration np^2. Thus for $M_L = 1$, $M_S = 1$ there is only one term ^3P on the left-hand side of eq. (7.60) and only one determinant $\{\overset{+}{1}\,\overset{+}{0}\}$ on the right-hand side, so

$$\zeta(np^2\ {}^3P) \times 1 \times 1 = \zeta(np) \times 1 \times \tfrac{1}{2} + \zeta(np) \times 0 \times \tfrac{1}{2},$$

or

$$\zeta(np^2\ {}^3P) = \tfrac{1}{2}\zeta(np). \tag{7.62}$$

7.3. Fine structure in *LS* coupling

In the example of $3p^2$ ^3P of Si I we quoted $\zeta(^3P) = 74 \cdot 02$ cm^{-1}. This was actually derived from the fitted *parameter* $\zeta(3p) = 148 \cdot 049$ cm^{-1}. If one were to attempt a calculation with $3p$ wave functions and a central potential $V(r)$, $\zeta(3p)$ would be the quantity one would obtain from eq. (7.56).

Let us take as a second example the $npn'p$ configuration. Now for $M_L = 1$, $M_S = 1$ there are two terms ^3D and ^3P, and two determinants, $\{\overset{+}{1}\ \overset{+}{0}\}$ and $\{\overset{+}{0}\ \overset{+}{1}\}$, in the sums of eq. (7.61):

$$\zeta(npn'p\ ^3D) + \zeta(npn'p\ ^3P) = \zeta(np) \times 1 \times \tfrac{1}{2} + \zeta(n'p) \times 0 \times \tfrac{1}{2}$$
$$+ \zeta(np) \times 0 \times \tfrac{1}{2} + \zeta(n'p) \times 1 \times \tfrac{1}{2}$$

or

$$\zeta(npn'p\ ^3D) + \zeta(npn'p\ ^3P) = \tfrac{1}{2}\zeta(np) + \tfrac{1}{2}\zeta(n'p). \tag{7.63}$$

We can get further than this by considering $M_L = 2$, $M_S = 1$, for which there is one term ^3D and one determinant $\{\overset{+}{1}\ \overset{+}{1}\}$. Thus

$$2\zeta(npn'p\ ^3D) = \tfrac{1}{2}\zeta(np) + \tfrac{1}{2}\zeta(n'p). \tag{7.64}$$

Therefore from eq. (7.63) and (7.64)

$$\zeta(npn'p\ ^3P) = \zeta(npn'p\ ^3D) = \tfrac{1}{4}\zeta(np) + \tfrac{1}{4}\zeta(n'p). \tag{7.65}$$

For shells more than half-full of equivalent electrons, nl^x, we have already remarked that the allowed terms are the same as for the configuration nl^{N-x} where $N = 2(2l + 1)$. The fine-structure parameters for terms of these two configurations are the same in magnitude, but opposite in sign:

$$\zeta(nl^x LS) = -\zeta(nl^{N-x}LS). \tag{7.66}$$

For example in the np^2 configuration the level $^3P_0$ is lowest, but for np^4 the fine structure is inverted and the $^3P_2$ level is the lowest level.

Because of this symmetry about the half-filled shell it follows that the fine structure vanishes, in first order, for *all terms* of a half-filled shell in *LS* coupling:

$$\zeta(nl^{N/2}LS) = 0. \tag{7.67}$$

One finds in the tables of energy levels that experimental fine-structure splittings are listed for half-filled shell configurations, but these are small and are usually due to breakdown of the coupling scheme.

Before we leave this section let us treat two cases of fine structure which we have left out of the discussion so far: the alkalis, and helium.

In chapter 6 we have discussed the gross structure of the alkalis as the problem of one electron in a central field, which is not a Coulomb field because of the electrostatic screening by the electrons of the core. The spin–orbit interaction is the perturbation (eq. (7.37)) for the single valence electron:

$$\mathscr{H}_2 = \xi(r)\mathbf{l} \cdot \mathbf{s}; \qquad \xi(r) = \frac{\hbar^2}{2m^2c^2} \cdot \frac{1}{r} \cdot \frac{dV}{dr}. \qquad (7.68)$$

The first-order energy shift is the expectation value of $\mathscr{H}_2$ in the $lsjm_j$ scheme as in eq. (4.25) for hydrogen:

$$\Delta E = \frac{\hbar^2}{2m^2c^2} \left\langle \frac{1}{r} \cdot \frac{dV}{dr} \mathbf{l} \cdot \mathbf{s} \right\rangle \qquad (7.69)$$

The angular part is

$$\langle lsjm_j| \mathbf{l} \cdot \mathbf{s} |lsjm_j \rangle = \tfrac{1}{2}\{j(j + 1) - l(l + 1) - s(s + 1)\} \qquad (7.70)$$

and the radial integral, eq. (7.56), can be written, in analogy with eq. (4.26),

$$\zeta(nl) = \frac{\hbar^2}{2m^2c^2} \frac{Z_{\text{eff}} \, e^2}{4\pi\varepsilon_0} \langle r^{-3} \rangle. \qquad (7.71)$$

$Z_{\text{eff}}\langle r^{-3} \rangle$ is often treated as a parameter to be evaluated from experiment. Thus eq. (7.69) becomes

$$\Delta E = \tfrac{1}{2}\zeta(nl)\{j(j + 1) - l(l + 1) - \tfrac{3}{4}\} \qquad (7.72)$$

since $s = \tfrac{1}{2}$, or in terms of the Bohr magneton

$$\Delta E = (\mu_0/4\pi)2\mu_B^2 Z_{\text{eff}}\langle r^{-3} \rangle \tfrac{1}{2}\{j(j + 1) - l(l + 1) - \tfrac{3}{4}\}. \qquad (7.73)$$

^2S terms are not split, having the value $j = \tfrac{1}{2}$ only, but all other terms are doublets with a separation between the $j = l + \tfrac{1}{2}$ and $j = l - \tfrac{1}{2}$ levels given by

$$\Delta W = (\mu_0/4\pi)2\mu_B^2 Z_{\text{eff}}\langle r^{-3} \rangle(l + \tfrac{1}{2}). \qquad (7.74)$$

If the central field were a pure Coulomb field $Z_{\text{eff}}\langle r^{-3} \rangle$ would be

$$\frac{Z^4}{a_0^3 n^3 l(l + \tfrac{1}{2})(l + 1)}$$

from table 2.3. To take account of the penetration of the core Landé modified this expression by replacing Z^4 by $Z_i^2 Z_0^2$, and n^3 by n^{*3}. Z_0 and n^* are the quantities appearing in the gross structure of the alkalis, eqs. (6.32) and (6.33), and Z_i is an effective 'inner' charge which must be treated as a parameter. As a very rough guide $Z_i \approx Z - 4$ for p-electrons and $Z_i \approx Z - 11$ for d-electrons. The Landé formula for the doublet splitting is therefore

$$\Delta W = \alpha^2 \frac{Z_i^2 Z_0^2}{n^{*3} l(l + 1)} \text{ Rydbergs.} \qquad (7.75)$$

7.3. Fine structure in LS coupling

The splitting of the lowest ^{2}P term in the alkalis increases with Z: $\Delta W(3p, \text{Na}) = 17{\cdot}2$ cm^{-1}; $\Delta W(4p, \text{K}) = 57{\cdot}7$ cm^{-1}; $\Delta W(5p, \text{Rb}) = 237{\cdot}6$ cm^{-1}; $\Delta W(6p, \text{Cs}) = 554{\cdot}1$ cm^{-1}. For a given element ΔW also falls off rapidly with increasing n and l. There are, however, anomalies in the magnitude of the fine-structure splitting in the alkalis which we shall not discuss.

The second example we want to discuss is the fine structure of the $1snp$ ^{3}P terms of helium: in particular we shall refer to the $1s2p$ ^{3}P term. Next to hydrogen, helium is the lightest element and the spin–orbit interaction is very small. Just as in hydrogen, one might expect that other contributions to the fine-structure splitting are comparable with the spin–orbit interaction and this is indeed the case.

As a non-relativistic approximation to the two-electron problem one can formulate the following magnetic interactions (already mentioned in chapter 5) which arise from relativistic effects: orbit–orbit, spin–other-orbit, and spin–spin in addition to the spin–orbit (i.e., spin–own-orbit) interaction already considered. Of these the orbit–orbit interaction does not contribute to a splitting of the term, only to a shift, so it can be ignored for the purpose of considering fine structure. In LS coupling the spin–other-orbit interaction can be written in an effective form $\zeta'(\gamma LS)\mathbf{L} \cdot \mathbf{S}$ just like the spin–orbit interaction. However, the parameter ζ' is different and in particular it does not depend on Z as strongly as the spin–orbit parameter (this partly accounts for the fact that it can be neglected in comparison with the spin–orbit interaction for large Z). Because of its angular form the spin–other-orbit interaction does not give a departure from the interval rule in LS coupling. The spin–spin interaction has quite a different form and does break down the interval rule. It may be written in the classical form of the interaction between two magnetic dipoles with magnetic moments proportional to $\mathbf{s}_1$ and $\mathbf{s}_2$ separated by a distance $\mathbf{r}_{12}$:

$$\frac{\mu_0}{4\pi} 4\mu_B^2 \left\{ \frac{\mathbf{s}_1 \cdot \mathbf{s}_2}{r_{12}^3} - 3 \frac{(\mathbf{s}_1 \cdot \mathbf{r}_{12})(\mathbf{s}_2 \cdot \mathbf{r}_{12})}{r_{12}^5} \right\}. \tag{7.76}$$

The effect of the three spin-dependent interactions in determining the fine structure of the $1s2p$ ^{3}P term of helium is shown in Fig. 7.5. The spin–orbit interaction gives a normal triplet obeying the interval rule. The addition of the spin–other-orbit interaction inverts the structure in this particular case but maintains the interval rule. The spin–spin interaction grossly distorts the structure, even putting the J-levels out of their normal order. The experimental splitting, which is only about 1 cm^{-1} overall, is shown on the right. The small disagreement between theory and experiment is attributed to the approximations made in dealing with these small effects.

The spin–spin interaction, so important here, is not the main cause of departures from the interval rule in heavier elements. It is relatively unimportant compared with breakdown of the LS coupling scheme due to the large size of the spin–orbit interaction. For example, the spin–spin interaction may often contribute only about $1\ \text{cm}^{-1}$ to a shift of the levels in a fine structure which extends over several thousand wavenumbers.

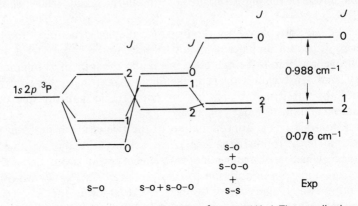

Fig. 7.5. The fine-structure splitting of the $1s2p$ ^{3}P term of He I. The contributions of the spin–orbit (s–o), spin–other-orbit (s–o-o), and spin–spin (s–s) interactions are shown cumulatively. The experimental splitting is on the right.

7.4. Relative intensities in *LS* coupling

Having discussed the structure of the energy levels in LS coupling we must now investigate what is observed experimentally, namely the spectral lines resulting from transitions between the energy levels. We need to know the selection rules for ΔL, ΔS, and ΔJ, particularly in electric dipole radiation. Relative intensities in single-photon transitions from one multiplet to another depend on the square of the matrix element of the multipole operator. For electric dipole radiation this is

$$|\langle \gamma SLJ| \sum_i e\mathbf{r}_i |\gamma'S'L'J'\rangle|^2. \qquad (7.77)$$

The easiest way to remember all selection rules is to apply a perfectly general rule which is soundly based in the theory of tensor operators. While that theory is too sophisticated for us to discuss it here we can nevertheless treat the rule as a mnemonic. We need to know the tensor rank, k, of the multipole operator and the space in which it operates. For example, the electric dipole operator of eq. (7.77) is a vector (a tensor of rank one; $k = 1$) and it operates in the total space of the electrons (label J), in orbital space, whether coupled (L) or not (l), but not in spin space (S). The electric quadrupole operator on the other hand is a tensor of rank two, $k = 2$, but it operates in the same spaces as the electric dipole operator.

The rule is: the tensor rank k of the multipole operator and the angular momentum quantum numbers of the two states involved in the transition must satisfy a triangular condition for each space in which the operator acts.

Thus for electric dipole radiation we have the following selection rules: $\sum_i e\mathbf{r}_i$ operates on the electrons, so $k = 1$, J, and J' must satisfy a triangular condition, hence

$$\Delta J = \pm 1, 0, \quad \text{but } J = 0 \nrightarrow J' = 0, \tag{7.78}$$

since the lengths 1, 0, 0 do not form a triangle. $\sum_i e\mathbf{r}_i$ does not operate in spin space, so $S' = S$ or

$$\Delta S = 0. \tag{7.79}$$

In L-space,

$$\Delta L = \pm 1, 0, \quad \text{but } L = 0 \nrightarrow L' = 0. \tag{7.80}$$

In the orbital space of single electrons the selection rule is

$$\Delta l_i = \pm 1, \qquad \Delta l_{j \neq i} = 0. \tag{7.81}$$

The transition $\Delta l_i = 0$ is not allowed because of parity considerations. Since $\sum_i e\mathbf{r}_i$ is odd under a parity transformation the parity of the configuration, $(-1)^{\sum_i l_i}$, must change in electric dipole radiation. The simultaneous jumping of two electrons is a rarer case which we shall not discuss. It occurs when there is strong configuration mixing, but even then the parities of the initial and final state must be different.

The selection rule for J is a strict one independent of the type of coupling of the orbital and spin angular momenta. But the rule, $\Delta S = 0$, for example, can be violated if the LS coupling is broken down by the spin–orbit interaction; then, of course, S does not strictly describe a constant of the motion for the stationary states. When S changes one speaks of intercombination lines. These are not uncommon. For example, the well-known mercury line 2,537 Å arises from the transition $6s^2 \, {}^1S_0$—$6s6p \, {}^3P_1$ from the lowest excited triplet to the $6s^2 \, {}^1S_0$ ground state. One expects to find, together with the appearance of this line which is forbidden in LS coupling, a departure from the interval rule in the ${}^3P$ term. This is the case. The spin–orbit interaction has matrix elements off-diagonal in S and L and it mixes the levels ${}^1P_1$ and ${}^3P_1$ of the $6s6p$ configuration; thus the ${}^3P$ term is not pure. The admixture of the ${}^1P_1$ level accounts for the non-zero intensity of the line 2,537 Å because the ${}^1S_0$—${}^1P_1$ transition is allowed in the zeroth-order approximation of LS coupling. In addition, the line is quite intense. The reason is that the intensity of a line is proportional not only to the transition probability but also to the population of the upper level, and the population of the ${}^3P_1$ level is high because of cascades from higher levels. The only mode of decay of this level by radiation is via the line 2,537 Å to the ground state. The transition $6s^2 \, {}^1S_0$—$6s6p \, {}^3P_0$ is not observed because its appearance

Mercury

would violate the strict selection rule for J (the spin–orbit interaction does not mix levels of different J). The $6s6p$ $^3P_0$ level is therefore metastable.

The members of a group of spectral lines which arise from the transition from one multiplet to another in LS coupling, and which satisfy the selection rules for electric dipole radiation, have relative intensities which are proportional to the square of the electric dipole matrix element given in eq. (7.77). The angular factor in this expression, depending on S, L, J, S', L', J', can be worked out once and for all, and tables of relative intensities have been compiled by White and Eliason.† The formulae themselves are quite complicated expressions in terms of the angular momentum quantum numbers and we shall not quote them here, but there are certain trends which are easy to remember and which help to give a feel for relative intensities. They are the following: (a) for a change ΔL ($\neq 0$) in the transitions from one term to another in LS coupling the strongest lines are those for which ΔJ has the same sign as ΔL. Of these, the intensities of the lines decrease as the magnitude of J decreases. The weakest lines are those for which ΔJ has the opposite sign from ΔL. The lines with $\Delta J = 0$ are intermediate in intensity. (b) for a change $\Delta L = 0$ the strongest lines are those for which $\Delta J = 0$ and J is large. Again, the intensities decrease with decreasing J. For $\Delta J = \pm 1$, the lines $J \to J - 1$ and $J - 1 \to J$ are equal in intensity, as would be expected from the symmetry of the situation, and are usually somewhat weaker than the strongest lines with $\Delta J = 0$.

There are also sum rules for intensities which we shall state shortly. But first let us consider an example of a group of transitions between fine-structure levels to illustrate not only the relative intensities but also the analysis of the observed frequencies and the deductions which can be drawn from such an analysis.

Let us consider a group of six lines lying close together near 4,450 Å in the spectrum of Ca I. Their wavenumbers are 22,432·3, 22,436·0, 22,441·6,

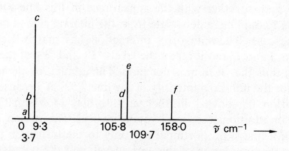

Fig. 7.6. A group of six lines in the spectrum of Ca I. The relative intensities, a—f, and the relative wavenumbers are shown.

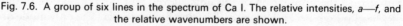

† These tables are reproduced in Kuhn and in Condon and Shortley for example.

7.4. Relative intensities in *LS* coupling

22,538·1, 22,541·8, 22,590·3 cm^{-1}. The wavenumbers of these six lines relative to that of the lowest frequency line, together with their intensities, marked a, b, c, d, e, f are shown in Fig. 7.6. In order to identify the energy levels connected by this group of transitions we can examine the frequency differences and the intensities. Since experimental measurement of intensities is much less reliable than measurement of frequency we shall concentrate on the frequencies and conclude what we can from them, using the intensities only as confirmation of the results. We know that calcium has two valence electrons, giving singlet and triplet terms. We assume that *LS* coupling is valid and that the six lines arise from transitions between two triplet terms ($\Delta S = 0$). We therefore look for an interval rule which will show up as the ratio of small integral numbers between some of the frequency differences. We find that

$$\frac{9 \cdot 3 - 3 \cdot 7}{3 \cdot 7 - 0} = \frac{5 \cdot 6}{3 \cdot 7} \approx \frac{3}{2};$$

and

$$\frac{105 \cdot 8 - 0}{158 \cdot 0 - 105 \cdot 8} = \frac{105 \cdot 8}{52 \cdot 2} \approx \frac{2}{1}.$$

It is also helpful to note that the difference 3·7 cm^{-1} appears twice: $(3 \cdot 7 - 0)$ and $(109 \cdot 5 - 105 \cdot 8)$, an example of the Rydberg–Ritz combination principle. We therefore suspect that, according to the interval rule, the levels $J = 3, 2, 1$ occur in one term, which must therefore be a ^{3}D term, and the levels $J = 2, 1, 0$ occur in the other, which must be a ^{3}P term. On this information we can now construct an energy level diagram, assuming normal multiplets, with transitions obeying the selection rules $\Delta L = -1 \, (^3\text{P} - {}^3\text{D})$ and $\Delta J = 0, \pm 1$. This diagram is shown in Fig. 7.7. The theoretical relative intensities are indicated below the energy-level diagram.

The configurations and terms involved in the group of lines are actually $4s4d \, ^3$D and $4s4p \, ^3$P. From the fine structure of the terms, in which the interval rule is quite well obeyed we conclude that

$$\zeta(4s4d \, ^3\text{D}) = 1 \cdot 9 \text{ cm}^{-1},$$
$$\zeta(4s4p \, ^3\text{P}) = 52 \cdot 8 \text{ cm}^{-1}.$$

We can go further and use eq. (7.61) to find the $\zeta(n_i l_i)$. Since $|4s4d \, ^3\text{D}$; $M_L = 2$; $M_S = 1\rangle = \{\overset{+}{0} \, \overset{+}{2}\}$, we obtain

$$2\zeta(4s4d \, ^3\text{D}) = 2 \times \tfrac{1}{2} \times \zeta(4d)$$

or

$$\zeta(4s4d \, ^3\text{D}) = \tfrac{1}{2}\zeta(4d),$$

137

and since

$$|4s4p\ ^3\mathrm{P};\ M_L = 1;\ M_S = 1\rangle = \{\overset{+}{0}\ \overset{+}{1}\},$$
$$\zeta(4s4p\ ^3\mathrm{P}) = \tfrac{1}{2}\zeta(4p).$$

Hence $\zeta(4d) = 3{\cdot}8\ \mathrm{cm}^{-1}$ and $\zeta(4p) = 105{\cdot}6\ \mathrm{cm}^{-1}$.

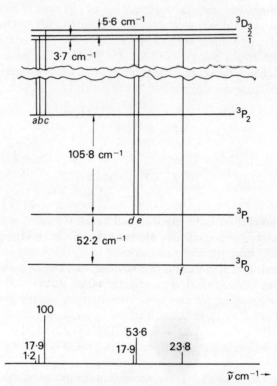

Fig. 7.7 The structure of the $4s4d\ ^3$D and $4s4p\ ^3$P terms in Ca I, with the transitions between them.

Now let us consider the relative intensities. We confirm that the strongest line, for $\Delta L = -1$, is the one with $\Delta J = -1$ involving the largest J: $^3\mathrm{P}_2$—$^3\mathrm{D}_3$. Also the other two lines with $\Delta J = -1$, $^3\mathrm{P}_1$—$^3\mathrm{D}_2$ and $^3\mathrm{P}_0$—$^3\mathrm{D}_1$, are less intense the smaller J is. The weakest line is that for which $\Delta J = +1$ where the sign of ΔJ is opposite to that of ΔL: the line $^3\mathrm{P}_2$—$^3\mathrm{D}_1$. In fact this line is so weak that it might be difficult to detect. The lines $^3\mathrm{P}_2$—$^3\mathrm{D}_2$ and $^3\mathrm{P}_1$—$^3\mathrm{D}_1$ are intermediate in intensity.

The sum rule which relates the relative intensities in transitions between two multiplets in LS coupling is the so-called Ornstein–Burger–Dorgelo sum rule: the sum of the intensities of all the transitions from an initial level or to a final level of a multiplet is proportional to the statistical weight, $2J + 1$, of that level; the constant of proportionality is common to all levels of a given multiplet. Thus in our example of the transitions ^3P—^3D the sums from the initial levels of the term ^3D give:

$$J = 3: \qquad\qquad c = 7k,$$
$$J = 2: \qquad\qquad b + e = 5k,$$
$$J = 1: \qquad\qquad a + d + f = 3k;$$

and the sums to the final levels of the term ^3P give

$$J' = 2: \qquad\qquad a + b + c = 5k',$$
$$J' = 1: \qquad\qquad d + e = 3k',$$
$$J' = 0: \qquad\qquad f = 1k'.$$

Taking ratios within a multiplet to eliminate the constants of proportionality k and k', and fixing arbitrarily one intensity on a relative scale, say $c = 100$, one obtains four equations and five unknowns, so the sum rule by itself is insufficient to determine the relative intensities in this case, and one would have to know at least one ratio of electric dipole matrix elements. But conversely, one can easily verify that the relative intensities given in Fig. 7.7 do satisfy the sum rules.

In a simpler case, such as ^2P—^2D, one can find the relative intensities by the sum rule alone. Referring to Fig. 7.8 we see that

$$\frac{q}{p + r} = \frac{6}{4} \quad \text{for } ^2D,$$

$$\frac{p + q}{r} = \frac{4}{2} \quad \text{for } ^2P.$$

Hence

$$q\!:\!r\!:\!p = 9\!:\!5\!:\!1.$$

Returning to the example of the transitions between the terms ^3D and ^3P we notice that the fine structure of the ^3D term is much less than that of the ^3P term. In the limit when the structure of the ^3D term is completely unresolved there remain only three lines in the spectrum with intensities in the ratios $5\!:\!3\!:\!1$. These numbers are just the statistical weights of the levels of the ^3P term. In the limit when the structure of the ^3P term is also completely unresolved there is just one line whose frequency is at the centre of gravity, or weighted mean, of the lines of the resolved spectrum. Since in *LS* coupling the perturbed levels are shifted from the position of the unperturbed term by

$$\Delta E = \tfrac{1}{2}\zeta\{J(J + 1) - L(L + 1) - S(S + 1)\}$$

(eq. (7.52)), the identity

$$\frac{1}{(2L + 1)(2S + 1)}\,\zeta\,\sum_{J=|L-S|}^{L+S}(2J + 1)\tfrac{1}{2}\{J(J + 1) - L(L + 1) \\ - S(S + 1)\} = 0 \quad (7.82)$$

simply states that the weighted mean of the energies of the levels belonging to a term coincides with the energy of the unperturbed term. The use of this identify together with a procedure of taking moments (intensity times displacement) about the centre of gravity of a group of lines corresponds exactly to the use of the sum rule for intensities in LS coupling (see problem 7.9).

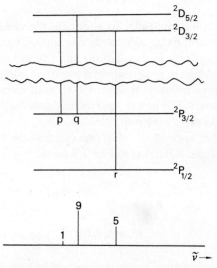

Fig. 7.8. Transitions between levels of the terms ^2D and ^2P, with relative intensities $q{:}r{:}p = 9{:}5{:}1$.

7.5. The *j–j* coupling approximation and other types of coupling

We have discussed, for a central field, the approximation in which the residual electrostatic interaction between electrons is large compared with the spin–orbit interaction. This approximation is called LS coupling because $\mathbf{L} = \sum_i \mathbf{l}_i$ and $\mathbf{S} = \sum_i \mathbf{s}_i$ are constants of the motion when the interaction between spin and orbit can be neglected.

We now turn to the other extreme in which, in a central field, the spin–orbit interaction is large compared with the residual electrostatic interaction. This case occurs in some configurations of heavy elements. It is

140

7.5. The *j–j* coupling approximation and other types of coupling

called *j–j* coupling for the following reason: if the residual electrostatic interaction between electrons is neglected, the electrons move *quite independently of each other* in a central field, each electron separately being subject to a spin–orbit interaction. Therefore the representation $(l_i s_i j_i m_{j_i})$ is appropriate just as in a one-electron atom, where $\mathbf{j}_i = \mathbf{l}_i + \mathbf{s}_i$. $\mathbf{l}_i$ and $\mathbf{s}_i$ couple up to form a resultant $\mathbf{j}_i$ which is a constant of the motion because there are no torques on one electron due to the others (the only effect of the other electrons is a contribution to the central field, a screening effect: it is the non-central interaction with other electrons which is being neglected). Thus the vector model pictures a rapid precession of each $\mathbf{l}_i$ and $\mathbf{s}_i$ about their resultant $\mathbf{j}_i$. The residual electrostatic interaction applied afterwards as a small perturbation causes a much slower precession of the $\mathbf{j}_i$ about their resultant $\mathbf{J}$ which is a constant of the motion in both the *LS* and *j–j* coupling schemes (see Fig. 7.9). But in the *j–j* coupling scheme *L* and *S* have no meaning.

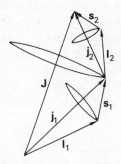

Fig. 7.9. Vector model for the angular momenta of two electrons in *j–j* coupling.

Let us consider the energy shifts caused by the spin–orbit interaction in *j–j* coupling. We start with a pure configuration of independent electrons and we apply the spin–orbit interaction

$$\mathcal{H}_2 = \sum_i \xi(r_i)\mathbf{l}_i \cdot \mathbf{s}_i \tag{7.83}$$

in first-order perturbation theory. The energy shift is the diagonal matrix element of $\mathcal{H}_2$ in the $(n_i l_i s_i j_i J M_J)$ representation, which is the representation (e) of section 7.1 (the representation (d), $(n_i l_i s_i j_i m_{j_i})$, would do equally well because $\mathcal{H}_2$ commutes with $\mathbf{l}_i^2$, $\mathbf{s}_i^2$, $\mathbf{j}_i^2$ and j_{z_i} as well as with $\mathbf{J}^2$ and J_z):

$$\Delta E = \langle n_i l_i s_i j_i J M_J| \sum_i \xi(r_i)\mathbf{l}_i \cdot \mathbf{s}_i |n_i l_i s_i j_i J M_J\rangle = \sum_i \Delta E_i, \tag{7.84}$$

where

$$\Delta E_i = \tfrac{1}{2}\zeta(n_i l_i)\{j_i(j_i + 1) - l_i(l_i + 1) - s_i(s_i + 1)\}. \tag{7.85}$$

141

Equation (7.84) just expresses the fact that the energy shift is the sum of the shifts for each independent electron. The levels remain degenerate with respect to J and M_J. The degeneracy with respect to J is lifted by the small electrostatic interaction between electrons; in applying this perturbation in first order the representation $(n_i l_i s_i j_i J M_J)$ is appropriate but $(n_i l_i s_i j_i m_{j_i})$ is not because $\sum_{i>j} e^2/4\pi\varepsilon_0 r_{ij}$ does not commute with j_{z_i}.

The notation in j–j coupling is one which names, in a two-electron configuration for example, l_1, l_2, j_1, j_2 and J. Thus for the $npn's$ configuration a level is denoted by $(p_{j_1}, s_{j_2})_J$ for which the possibilities are $j_i = l_i \pm \frac{1}{2}$ $(l_i \neq 0)$, $J = j_1 + j_2, \ldots, |j_1 - j_2|$: $(p_{1/2}, s_{1/2})_1$, $(p_{1/2}, s_{1/2})_0$, $(p_{3/2}, s_{1/2})_2$, and $(p_{3/2}, s_{1/2})_1$. The number of levels, four in this example, is the same in j–j coupling as in LS coupling, with the same four values of J which is a constant of the motion in both schemes. In LS coupling we have for the configuration $npn's$ the four levels $^1P_1$, $^3P_0$, $^3P_1$, $^3P_2$.

Let us now consider an example of two valence electrons in j–j coupling. The configuration $np(n+1)s$ in heavy elements is one which is often quoted, for example the $5p6s$ configuration of Sn I ($Z = 50$). Here we have to deal with $\zeta(5p)$ because the s-electron does not have spin–orbit splitting. $\zeta(5p)$ which is proportional to $\langle 1/r \cdot dV/dr \rangle$, or $Z_{eff} \langle r^{-3} \rangle$, for the $5p$ electron is large, about 2,700 cm^{-1}—the splitting is about 4,000 cm^{-1} (see Fig. 7.10). The loosely bound $6s$ electron hardly affects the magnitude of $Z_{eff} \langle r^{-3} \rangle_{5p}$ at all, for when it is removed the spin–orbit splitting in the ground term $5p \ ^2P$ of Sn II is almost the same—4,251 cm^{-1}

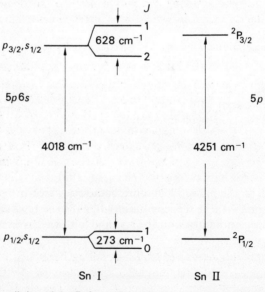

Fig. 7.10. The splitting of the $5p6s$ configuration in Sn I as an example of j–j coupling compared with the splitting of the $5p$ configuration in Sn II.

7.5. The *j–j* coupling approximation and other types of coupling

—as in the configuration $5p6s$ of Sn I. The same kind of comparison can be made for Bi II $6p7s$ and Bi III $6p$ ^{2}P, for example, where the spin–orbit interaction for the $6p$ electron gives a splitting in both cases of about 20,000 cm^{-1} (notice the very large magnitudes of spin–orbit interaction which we are talking about in *j–j* coupling).

The residual electrostatic interaction in $5p6s$ of Sn I accounts for the splitting of only 273 cm^{-1} between $(p_{1/2}, s_{1/2})_0$ and $(p_{1/2}, s_{1/2})_1$, and of 628 cm^{-1} between $(p_{3/2}, s_{1/2})_2$ and $(p_{3/2}, s_{1/2})_1$ (see Fig. 7.10). The electrostatic splitting in an $npn's$ configuration arises only from an exchange effect as in eqs. (7.11) and (7.12) in which the Slater exchange integral G_1 determines the splitting (F_0 is common to all levels of the configuration). One concludes from the experimental data for our example that the overlap between the $5p$ and $6s$ wave functions is already small, and one would expect that it would become even smaller as n', the principal quantum number of the s-electron, increases and hence that the *j–j* coupling scheme would become a better and better approximation. In view of the absence of spin–orbit interaction for the $n's$ electron perhaps it would be better to refer to the coupling as *j–s* coupling in this case rather than *j–j* coupling.

Along the sequence $np(n+1)s$ from light to heavy elements, for example C($2p3s$), Si($3p4s$), Ge($4p5s$), Sn($5p6s$), there is a progression from *LS* to *j–j* coupling. Ge($4p5s$) is an example of *intermediate* coupling for which neither L, S nor j_1, j_2 are even approximately good quantum numbers. That is to say, the two perturbations, residual electrostatic and spin–orbit, are of the same order of magnitude. To calculate the energy levels in intermediate coupling one would have to go to the trouble of solving a secular equation in which both interactions are treated on an equal footing.

In the lower-lying configurations of neutral atoms the *j–j* coupling approximation is not often found, in other words the conditions for small electrostatic interaction are not often satisfied, even in heavy elements. For example the $5s5p$ configuration of Cd I ($Z = 48$) gives rise to quite a different situation from that of $5p6s$ in Sn I ($Z = 50$). In Cd the value of $\zeta(5p)$ is of the order of 1100 cm^{-1}, compared with $\zeta(5p) \sim 2{,}700$ cm^{-1} in Sn, but the electrostatic splitting between the ^{1}P and ^{3}P terms is about 13,000 cm^{-1}, so the *LS* coupling approximation is much more appropriate.

Other types of coupling scheme are sometimes met in special cases. One such case is found in the rare gases when the core has one hole and there is one electron which is on the average far away from the core, for example, the configuration $3p^5 4f$ in argon. The $3p^5$ configuration by itself is appropriately described in *LS* coupling by L_1, S_1, and J_1: the levels are ^{2}P$_{1/2, 3/2}$, and the spin–orbit interaction of the hole gives a splitting of about 1,400 cm^{-1} between J_1 levels. The electrostatic interaction between the distant $4f$ electron and the $3p$ electrons is too weak to destroy J_1 as a good quantum number, so we retain J_1 and form a new quantum number K, where $\mathbf{K} = \mathbf{J}_1 + \mathbf{l}$ and $\mathbf{l}$ is the orbital angular momentum of the outer electron (the $4f$ electron in this case). This small electrostatic interaction accounts for a splitting of the order of 20 cm^{-1} between different K levels for a given J_1: $K = J_1 + l, J_1 + l - 1, \ldots,$ $|J_1 - l|$. Finally the spin–orbit interaction for the $4f$ electron is weakest of all, splitting each (J_1, K) level into two J levels: $J = K \pm \frac{1}{2}$. This splitting is less than 0·5 cm^{-1}. When the various interactions differ by orders of magnitude in this particular way one speaks of J_1–l coupling for which the following notation has been devised: first one writes the $L_1 S_1 J_1$ for the core, e.g., ($^2P_{3/2}$), then nl of the valence electron, $4f$, then K, the resultant of J_1 and l, in square brackets with the total angular momentum of the atom, J, as a subscript. Thus: $(^2P_{3/2})4f[3\frac{1}{2}]_3$.

In closing this chapter we should repeat that the LS coupling approximation is by far the most common basis for a description of the simple atomic spectra.

Problems

(Those problems marked with an asterisk are more advanced.)

7.1. Show that $\sum_{J=|L-S|}^{L+S} (2J + 1) = (2L + 1)(2S + 1)$, and hence that the number of non-degenerate states of a term is the same in the $(LSJM_J)$ representation as in the (LSM_LM_S) representation.

7.2. Use the branching rules to write out the terms of the $npn'dn''p$ configuration. Note that terms of a given L, S occur more than once. Distinguish these by naming their parents.

7.3. By using the rule that for the terms of two equivalent electrons $S + L$ must be even, write out the allowed terms of the configurations nd^2, nf^2, ng^2.

*7.4. It is a property of determinantal product functions that matrix elements, between such functions, of a sum of single-particle operators $\sum_i f_i$, are just the same as if simple product functions had been used. It is worth verifying this by brute force for a simple case, say a 2×2 determinant

$$2^{-1/2} \begin{vmatrix} u_a(1) & u_a(2) \\ u_b(1) & u_b(2) \end{vmatrix}$$

and an operator $f_1 + f_2$.

Hence verify for a p^6 configuration that in the central-field approximation $\langle \sum_i \xi(r_i) \mathbf{l}_i \cdot \mathbf{s}_i \rangle$ vanishes when the sum is taken over all electrons of a closed shell.

*7.5. For the np^2 configuration show that, in terms of the determinantal product states with $m_{l_1} + m_{l_2} = 0$ and $m_{s_1} + m_{s_2} = 0$ the LS coupled states with $M_L = 0$ and $\bar{M}_S = 0$ are

$$|^1D; 0, 0\rangle = (6)^{-1/2}\{\overset{+}{1}\ \overset{-}{-1}\} + (6)^{-1/2}\{\overset{+}{-1}\ \overset{-}{1}\} + 2(6)^{-1/2}\{\overset{+}{0}\ \overset{-}{0}\},$$

$$|^3P; 0, 0\rangle = (2)^{-1/2}\{\overset{+}{1}\ \overset{-}{-1}\} - (2)^{-1/2}\{\overset{+}{-1}\ \overset{-}{1}\},$$

$$|^1S; 0, 0\rangle = (3)^{-1/2}\{\overset{+}{1}\ \overset{-}{-1}\} + (3)^{-1/2}\{\overset{+}{-1}\ \overset{-}{1}\} - (3)^{-1/2}\{\overset{+}{0}\ \overset{-}{0}\}.$$

7.6. The Landé formula for the doublet splitting in the alkalis is given in eq. (7.75). From the following data for the lowest ^2P terms work backwards to evaluate Z_i and hence test the empirical formula $Z_i = Z - 4$ for p-electrons:

Element	Na	K	Rb	Cs
n	3	4	5	6
$\Delta W(np)(\text{cm}^{-1})$	17·2	57·7	237·6	554·1
$\delta(p)$	0·884	1·767	2·714	3·623

7.7. By considering classically the potential energy of interaction of two magnetic dipoles $\boldsymbol{\mu}_1 = 2\mu_B\mathbf{s}_1$ and $\boldsymbol{\mu}_2 = 2\mu_B\mathbf{s}_2$ separated by a distance r_{12}, show that the magnetic spin–spin interaction between two electrons in an atom can be written

$$(\mu_0/4\pi)4\mu_B^2\left\{\frac{\mathbf{s}_1\cdot\mathbf{s}_2}{r_{12}^3} - 3\frac{(\mathbf{s}_1\cdot\mathbf{r}_{12})(\mathbf{s}_2\cdot\mathbf{r}_{12})}{r_{12}^5}\right\}.$$

7.8. Write down the selection rules for electric quadrupole radiation in the LS coupling approximation.

7.9. Consider the transition ^2P—^2D in LS coupling, and let the intensities of the fine-structure components be p, q, and r as in Fig. 7.8. By taking moments about the centre of gravity of this group of components and using eq. (7.82), evaluate p, q in terms of r and show that this procedure is equivalent to the use of the sum rule for intensities.

*7.10. What are the allowed terms of the configuration np^3? Use the diagonal sum rule, eq. (7.61), to show by brute force that $\zeta(np^3LS) = 0$ for each of the terms of this half-filled shell configuration as stated in eq. (7.67).

*7.11. What are the allowed terms of the configuration nd^3? Note that this is the simplest configuration of equivalent electrons for which a term of given L and S occurs more than once.

8. Interaction with static external fields

In this chapter we consider the effect of applying a static magnetic or electric field along the z-axis. Such a field establishes the z-axis as a preferred axis in space, and so we expect the degeneracy in magnetic quantum number to be lifted, or partially lifted.

8.1. Zeeman effect in *LS* coupling

We consider first the interaction of the atomic electrons with an external magnetic field **B**. This gives rise to the Zeeman effect. We treat the interaction as a small perturbation, written in the classical form

$$\mathscr{H}_M = -\mathbf{\mu} \cdot \mathbf{B} \tag{8.1}$$

where $\mathbf{\mu}$ is the total magnetic moment of the electrons. From eqs. (4.4) and (4.12) $\mathbf{\mu}$ is written in terms of the orbital and spin magnetic moments:

$$\mathscr{H}_M = \left(\sum_i \mu_B \mathbf{l}_i + \sum_i g_s \mu_B \mathbf{s}_i\right) \cdot \mathbf{B} \tag{8.2}$$

where the orbital and spin g-factors, $g_l \equiv 1$ and $g_s \approx 2$, are defined here to be positive numbers. Thus the sign of μ_l with respect to $\mathbf{l}$, $\mathbf{\mu}_l = -\mu_B\mathbf{l}$, and of $\mathbf{\mu}_s$ with respect to $\mathbf{s}$, $\mathbf{\mu}_s = -g_s\mu_B\mathbf{s}$, is kept explicit and is not incorporated in the g-factor.

It is important to consider the size of the perturbation $\mathscr{H}_M$ compared with other terms in the Hamiltonian. We assume that the *LS* coupling approximation is valid, so that the zeroth-order Hamiltonian contains the central field, eq. (7.1), and the residual electrostatic interaction, eq. (7.2), which is assumed to be large compared with the spin–orbit interaction, eq. (7.3). We also include the spin–orbit interaction in the zeroth-order Hamiltonian, thereby making the assumption that it is large compared with the Zeeman interaction. This approximation is the *weak-field* case, meaning that the energy splitting, $\sim \mu_B B$, produced by the external field is small compared with the fine structure, $\sim \zeta(LS) \sim \mu_B B_{\text{int}}$, or that B is weak compared with the magnetic field B_{int} internal to the

8.1. Zeeman effect in *LS* Coupling

atom. The criterion $B \ll \zeta/\mu_B$ for a weak field leads to a value of B for a typical case of $\zeta \sim 100\ \text{cm}^{-1}$: $B \ll B_{\text{int}} \sim 100\ \text{T}$. Thus it is not very often that the weak-field condition will be violated. Put another way round, commonly attainable laboratory fields of $B \sim 1\ \text{T}$ will be weak if $\zeta \gg \mu_B B \sim 1\ \text{cm}^{-1}$.

In weak field, therefore, we consider a fine-structure level labelled by (γLSJ), that is, the level J is isolated from other levels $(\gamma LSJ')$. In first-order perturbation theory the Zeeman energy shift is

$$\Delta E = \langle \gamma LSJM_J | \mathcal{H}_M | \gamma LSJM_J \rangle. \tag{8.3}$$

We may write $\mathcal{H}_M$ in the effective form

$$\begin{aligned}
\mathcal{H}_M &= \mu_B(\mathbf{L} + g_s\mathbf{S}) \cdot \mathbf{B} \\
&= \mu_B B(L_z + g_s S_z), \tag{8.4}
\end{aligned}$$

which is an appropriate form for *LS* coupling. Although the zeroth-order eigenfunctions of eq. (8.3) are degenerate in M_J, we can proceed as in non-degenerate perturbation theory because $\mathcal{H}_M$ commutes with $J_z = L_z + S_z$.

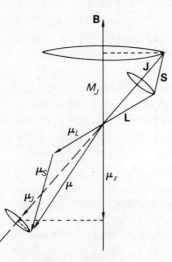

Fig. 8.1. Vector model for $\mathbf{\mu}$, showing the projections first on the direction of $\mathbf{J}$ and then on the z-axis.

To evaluate the matrix element of L_z and of S_z we take the projection of $\mathbf{L}$ on $\mathbf{J}$ first and then project on to the z-axis in a manner analogous to eq. (7.49). The vector model (see Fig. 8.1) says that $\mathbf{L}$ and $\mathbf{S}$ are precessing rapidly about $\mathbf{J}$ with a rate of precession proportional to $\zeta(LS)$ and $\mathbf{J}$ is precessing slowly about the z-axis with a rate proportional to $\mu_B B$ ($\ll \zeta(LS)$).

Thus we write

$$\langle \gamma LSJM_J | \, L_z \, | \gamma LSJM_J \rangle = \langle \gamma LSJM_J | \, \frac{\mathbf{L} \cdot \mathbf{J}}{J(J+1)} \, J_z \, | \gamma LSJM_J \rangle, \quad (8.5)$$

and similarly for S_z. Now

$$\mathbf{S} \cdot \mathbf{S} = (\mathbf{J} - \mathbf{L}) \cdot (\mathbf{J} - \mathbf{L}) = \mathbf{J}^2 + \mathbf{L}^2 - 2\mathbf{L} \cdot \mathbf{J}, \quad (8.6)$$

therefore

$$\mathbf{L} \cdot \mathbf{J} = \tfrac{1}{2}(\mathbf{J}^2 + \mathbf{L}^2 - \mathbf{S}^2). \quad (8.7)$$

Similarly

$$\mathbf{S} \cdot \mathbf{J} = \tfrac{1}{2}(\mathbf{J}^2 - \mathbf{L}^2 + \mathbf{S}^2). \quad (8.8)$$

The energy shift becomes

$$\Delta E = \mu_B B \langle \gamma LSJM_J | \, \tfrac{1}{2}\{(\mathbf{J}^2 + \mathbf{L}^2 - \mathbf{S}^2)$$
$$+ g_s(\mathbf{J}^2 - \mathbf{L}^2 + \mathbf{S}^2)\} \, \frac{J_z}{J(J+1)} \, | \gamma LSJM_J \rangle$$
$$= \left\{ \frac{J(J+1) + L(L+1) - S(S+1)}{2J(J+1)} \right.$$
$$\left. + g_s \frac{J(J+1) - L(L+1) + S(S+1)}{2J(J+1)} \right\} \mu_B B M_J \quad (8.9)$$

because $|\gamma LSJM_J\rangle$ is a simultaneous eigenfunction of $\mathbf{L}^2$, $\mathbf{S}^2$, $\mathbf{J}^2$, and J_z. But going back to eq. (8.1) we can define an effective operator for the total magnetic moment $\mathbf{\mu}$. Since $\mathbf{\mu}$ is a vector operator and we require only its matrix element diagonal in J we can write, in analogy with equation (7.43),

$$\mathbf{\mu}_{\text{eff}} = \frac{(\mathbf{\mu} \cdot \mathbf{J})}{J(J+1)} \, \mathbf{J}, \quad (8.10)$$

which is equivalent to the vector model prescription for projecting $\mathbf{\mu}$ on to $\mathbf{J}$. The matrix element, diagonal in J, of the projection factor is independent of M_J, and we can therefore define an effective g-factor by

$$\mathbf{\mu}_{\text{eff}} = -g_J \mu_B \mathbf{J} \quad (8.11)$$

where

$$g_J = g_J(L, S, J). \quad (8.12)$$

The energy shift is

$$\Delta E = \langle \gamma LSJM_J | -\mu_z B | \gamma LSJM_J \rangle$$
$$= \langle \gamma LSJM_J | \, g_J \mu_B B J_z \, | \gamma LSJM_J \rangle$$
$$= g_J \mu_B B M_J. \quad (8.13)$$

8.1. Zeeman effect in *LS* coupling

Comparing eq. (8.13) with eq. (8.9) we find

$$g_J = \frac{J(J + 1) + L(L + 1) - S(S + 1)}{2J(J + 1)}$$
$$+ g_s \frac{J(J + 1) - L(L + 1) + S(S + 1)}{2J(J + 1)}. \quad (8.14)$$

g_J is called the Landé g-value. We see, according to the assumptions used to formulate it, that g_J has meaning only in the weak field case.

For singlets $S = 0$, and from eq. (8.14) $g_J = 1$, independent of L, S, and J; that is to say, in the absence of a resultant spin the Zeeman effect comes just from the interaction of the orbital magnetic moment of the electrons with the external field. Let us consider the splitting of a $^1D_2$ level as an example. For $J = 2$ the $2J + 1$ possible values of M_J are 2, 1, 0, -1, -2. According to eq. (8.13) the M_J degeneracy of a level is lifted and each level J splits into $2J + 1$ *states*, labelled by M_J. The field-dependence of the energy of each state, which is linear in B in first order, is shown in Fig. 8.2. From eq. (8.13) one can express the expectation value of μ_z as the negative slope of the energy with respect to field:

$$\langle \mu_z \rangle = -\frac{\partial E}{\partial B} = -g_J \mu_B M_J \quad (8.15)$$

where in the case of a singlet $g_J = 1$.

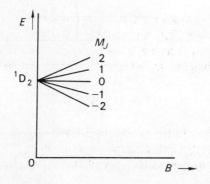

Fig. 8.2. The energies of the Zeeman states of a singlet level as a function of magnetic field **B**.

Now let us consider a transition between singlet levels in a weak magnetic field. The electric dipole selection rules for M_J are

$$\Delta M_J = 0, \pm 1; M_J = 0 \nrightarrow M'_J = 0 \text{ if } \Delta J = 0. \quad (8.16)$$

These are related to the polarization of the electric vector (see section 8.4): $\Delta M_J = 0$ corresponds to an electric dipole oscillating in the z-direction (π polarization) and $\Delta M_J = \pm 1$ corresponds to oscillation in the x–y plane (σ polarization). For a transition $^1P_1$—$^1D_2$, for example, at a given value of B the energy spacing between the states is $\mu_B B$, the same for each term. Therefore there are only three different frequencies symmetrically disposed about the zero-field frequency. They are given by

$$h\nu = h\nu_0 + \mu_B B \, \Delta M_J, \qquad (8.17)$$

where $h\nu_0$ is the difference in energy of the unperturbed levels. These transitions are shown in Fig. 8.3. All three lines, separated by $\mu_B B/h$, are

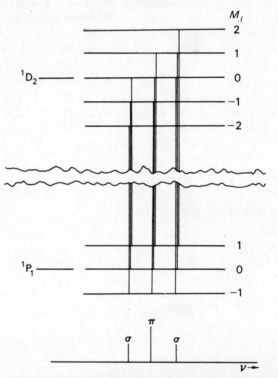

Fig. 8.3. The components of the transition $^1P_1$—$^1D_2$ in a magnetic field, illustrating the normal Zeeman effect in transverse observation.

observed when viewed transverse to the magnetic field: the π and σ components are plane polarized at right angles to each other. In longitudinal observation (along the field direction) the π component is missing and the σ components are circularly polarized in opposite directions. This so-called Lorentz triplet of one π and two σ lines is characteristic of the 'normal' Zeeman effect in electric dipole radiation. The appearance of this normal Zeeman effect in spectral lines was explained by Lorentz in classical terms before the introduction of electron spin—the Zeeman effect itself was discovered as early as 1896 (see problem 8.11).

8.1. Zeeman effect in *LS* coupling

For the general case of transitions between multiplets in *LS* coupling the situation is different. This 'anomalous' Zeeman effect depends on the fact that $g_s \neq 1$. For if g_s were 1 in eq. (8.14) we should have $g_J = 1$, independent of L, S, and J as in the normal Zeeman effect; in terms of the vector model, Fig. 8.1, μ would lie exactly along $\mathbf{J}$. However, this is not the case and the frequency of the transition $(\gamma'L'SJ'M'_J) \to (\gamma LSJM_J)$ is given by

$$h\nu = (E' + \Delta E') - (E + \Delta E)$$
$$= h\nu_0 + \mu_B B(g'_J M'_J - g_J M_J) \qquad (8.18)$$

from eq. (8.13).

Let us illustrate eq. (8.18) by reference to the sodium D lines: $^2S_{1/2}$—$^2P_{1/2}$ and $^2S_{1/2}$—$^2P_{3/2}$. The fine-structure splitting between the ^2P levels is $17 \cdot 2 \text{ cm}^{-1}$, so $\zeta(^2P)$ is about 11 cm^{-1}. A magnetic field is weak in this context if $B \ll \zeta(^2P)/\mu_B \sim 25 \text{ T}$. The g-factors are $g_J(^2S_{1/2}) = 2$; $g_J(^2P_{1/2}) = \frac{2}{3}$; $g_J(^2P_{3/2}) = \frac{4}{3}$ from eq. (8.14). Therefore the energy levels and transitions for a weak field are as shown in Fig. 8.4.

The observation of the Zeeman effect has been used as an important aid in spectral analysis. After a tentative assignment of L, S, and J to the multiplets involved in a group of transitions has been made (along the lines of the example ^3P—^3D given in section 7.4), the qualitative nature of the Zeeman patterns and a quantitative measurement of the values of g_J can be used to verify the assignment because g_J depends on L, S, and J. Furthermore, if it happens that no assignment of angular momentum quantum numbers has been made initially, not even a tentative one, as may well be the case in a very complicated spectrum, the Zeeman effect can be a very powerful tool in the identification of levels.

Many precise measurements of g_J, particularly for ground levels, have now been made by the methods of radiofrequency spectroscopy. Departures from the Landé g-value are caused by a breakdown of the *LS* coupling scheme, that is when the spin–orbit interaction is large enough to mix levels of different L and S (but the same J, because the spin–orbit interaction commutes with $\mathbf{J}^2$ and J_z). However, an experimental measurement of g_J for an impure level is not a particularly sensitive test of the amount of impurity. For example, if the level $|\gamma LSJ\rangle$ has admixed into it the level $|\gamma'L'S'J\rangle$ with amplitude a, the departure of the measured g_J from the Landé value $g_J(LSJ)$ is only $(g_J(LSJ) - g_J(L'S'J))a^2$. If a is small a measurement of g_J does not determine a very well. Configuration mixing by the electrostatic interaction does not by itself introduce any departure from the Landé value whatever because $\sum_{i>j} e^2/4\pi\varepsilon_0 r_{ij}$ commutes with $\mathbf{L}^2$, $\mathbf{S}^2$, and $\mathbf{J}^2$, and levels admixed in this way therefore have the same g_J.

Let us now turn to the opposite extreme in *LS* coupling: the strong-field case. The criterion for a magnetic field to be strong is $\mu_B B \gg \zeta(\gamma LS)$.

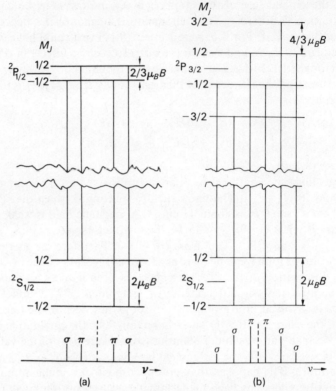

Fig. 8.4. The Zeeman effect in the D lines of sodium, in transverse observation: (a) $^2S_{1/2}$—$^2P_{1/2}$; (b) $^2S_{1/2}$—$^2P_{3/2}$.

In view of the magnitudes already quoted this situation is likely to be realized only in light elements for which $\zeta \ll 1 \text{ cm}^{-1}$.

For a strong field we have the following inequality for the Zeeman and spin–orbit interactions:

$$\mu_B B(L_z + g_s S_z) \gg \zeta(\gamma LS)\mathbf{L} \cdot \mathbf{S}. \tag{8.19}$$

Therefore we omit the spin–orbit interaction as a first approximation, that is we remove it from the zeroth-order Hamiltonian which now includes only the central field and residual electrostatic interaction. The appropriate zeroth-order wave function is $|\gamma LSM_L M_S\rangle$ as for a *term*. There is no interaction between spin and orbit, and on applying $\mathscr{H}_M$ as a perturbation in first order J has no meaning. According to the vector model (see Fig. 8.5) $\mathbf{L}$ and $\mathbf{S}$ precess independently about the direction of $\mathbf{B}$ with projections M_L and M_S on the z-axis (i.e., the direction of $\mathbf{B}$). There is degeneracy in M_L and M_S, but $\mathscr{H}_M$ commutes with L_z and S_z, so

$$\Delta E = \langle \gamma LSM_L M_S| \mu_B B(L_z + g_s S_z) |\gamma LSM_L M_S\rangle$$
$$= (M_L + g_s M_S)\mu_B B. \tag{8.20}$$

8.1. Zeeman effect in *LS* coupling

Having lifted the degeneracy in M_L and M_S we now apply the spin–orbit interaction $\zeta(\gamma LS)\mathbf{L} \cdot \mathbf{S}$ as a smaller perturbation. Notice that we are able, for once, to use *non-degenerate* perturbation theory in which we require, in first order, only the matrix element of $\zeta(\gamma LS)\mathbf{L} \cdot \mathbf{S}$ which is diagonal in the zeroth-order representation, even though the operator has non-vanishing off-diagonal elements in this representation. The diagonal matrix element is

$$\langle \gamma LSM_LM_S| \, \zeta(\gamma LS)\mathbf{L} \cdot \mathbf{S} \, |\gamma LSM_LM_S\rangle = \zeta(\gamma LS)M_LM_S \quad (8.21)$$

as in eq. (7.58) so in the lowest approximation the contribution to an energy shift of the Zeeman interaction and the spin–orbit interaction in a strong field is

$$\Delta E = (M_L + g_sM_S)\mu_B B + \zeta(\gamma LS)M_LM_S. \quad (8.22)$$

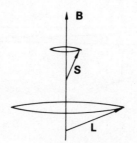

Fig. 8.5. Vector model showing **S** and **L** precessing independently about the direction of **B** in a strong magnetic field.

The electric dipole selection rules are

$$\Delta M_S = 0; \qquad \Delta M_L = 0 \quad \pi \text{ polarization},$$
$$= \pm 1 \quad \sigma \text{ polarization}, \quad (8.23)$$

so the frequencies in a transition from one term to another in strong field are

$$h\nu = h\nu_0 + \mu_B B\Delta M_L + \{\zeta(\gamma LS)M_SM_L - \zeta(\gamma'L'S)M_SM'_L\}. \quad (8.24)$$

This strong-field limit of the Zeeman effect is called the Paschen–Back effect. When ζ can be neglected eq. (8.24) is identical to the normal Zeeman effect for there is no dependence on the spin—S and L are uncoupled and $\Delta M_S = 0$.

The energies of the states of a ^2P term in strong field (eq. (8.22)) are illustrated on the right-hand side of Fig. 8.6. The states ($M_L = -1$, $M_S = \frac{1}{2}$) and ($M_L = 1$, $M_S = -\frac{1}{2}$) coincide if $g_s = 2$ exactly. At the left of Fig. 8.6 are shown the weak-field Zeeman states whose energy shifts from the fine-structure levels are given by eq. (8.13).

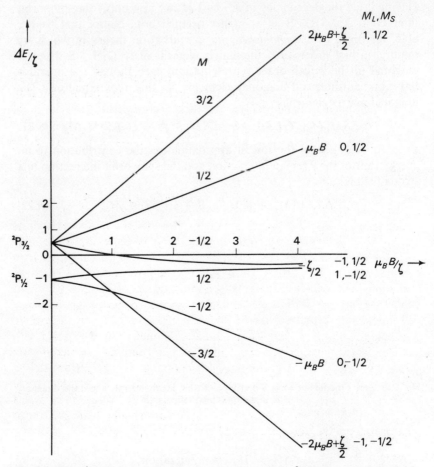

Fig. 8.6. The dependence of the Zeeman energy states of a ^{2}P term on magnetic field. $\Delta E/\zeta$ is plotted as a function of $\mu_B B/\zeta$. At the right are shown the energies ΔE in strong field, on the assumption that $g_s = 2$, together with the strong-field labels M_L, M_S.

In the intermediate field region, where the Zeeman interaction and the spin–orbit interaction are of the same order of magnitude, one has to solve a secular equation to find the energies of the states. We shall not do this here but we shall make certain observations about the intermediate field region. Neither J nor M_L, M_S are good quantum numbers because the combined Hamiltonian

$$\mathcal{H}' = (L_z + g_s S_z)\mu_B B + \zeta \mathbf{L} \cdot \mathbf{S} \qquad (8.25)$$

does not commute with $\mathbf{J}^2$ nor with L_z and S_z separately. But it does commute with $J_z = L_z + S_z$, so

$$M = M_J = M_L + M_S \qquad (8.26)$$

154

8.2. Quadratic Stark effect

is a good quantum number at all fields. Therefore $\mathcal{H}'$ does not have matrix elements off-diagonal in M. The matrix of $\mathcal{H}'$ which has to be diagonalized breaks up into submatrices of given M, of which there are four in our example: $M = \frac{3}{2}, \frac{1}{2}, -\frac{1}{2}, -\frac{3}{2}$.

If we pursue the low-field approximation to second-order perturbation theory in $\mathcal{H}_M$ we find that the M states, which depend linearly on B in first order, begin to bend quadratically in B according to

$$\Delta E^{(2)}(^2P_{3/2}, M) = \frac{|\langle ^2P_{3/2}, M| \, \mu_B B(L_z + g_s S_z) \,|^2P_{1/2}, M \rangle|^2}{E_0(^2P_{3/2}) - E_0(^2P_{1/2})}, \quad (8.27)$$

and

$$\Delta E^{(2)}(^2P_{1/2}, M) = -\Delta E^{(2)}(^2P_{3/2}, M). \quad (8.28)$$

The result is a repulsion between states of the same M: the states $(^2P_{3/2}, \frac{1}{2})$ and $(^2P_{1/2}, \frac{1}{2})$ bend away from each other. The generalization of this statement is that *states with the same M never cross* on an energy level diagram. This rule enables one to join up the low- and high-field states unambiguously, however complicated the diagram. Notice that the states $M = \pm\frac{3}{2}$ do not contribute to higher orders in perturbation theory because in eq. (8.27) the level $^2P_{1/2}$ does not have $|M| = \frac{3}{2}$. Therefore the energies of the $M = \pm\frac{3}{2}$ states are exactly linear in B at all fields.

We shall not discuss the Zeeman effect in j–j coupling: we merely note that, by taking the appropriate projections, it is straightforward to evaluate $g_J = g_J(l_1, s_1, j_1; l_2, s_2, j_2; J)$ in weak fields for two electrons, starting from the Zeeman Hamiltonian of eq. (8.2). 'Weak' field in this context means that the Zeeman splitting is much smaller than that due to the residual electrostatic interaction (see problem 8.2).

8.2. Quadratic Stark effect

When an atom is placed in an external electric field $\mathbf{E}$ whose direction defines the z-axis there is an additional term in the Hamiltonian:

$$\begin{aligned} \mathcal{H}_E &= -\sum_i (-e\mathbf{r}_i) \cdot \mathbf{E} \\ &= eE_z \sum_i z_i \\ &\equiv eE_z z. \quad (8.29) \end{aligned}$$

We consider first an atom in a state $|\gamma J M_J\rangle$ of well-defined parity, given by

$$(-1)^{\Sigma_i l_i}.$$

Then, as we have seen in eq. (3.80), the expectation value of $\mathcal{H}_E$ vanishes,

$$\langle \gamma J M_J| \, \mathcal{H}_E \,|\gamma J M_J\rangle = 0, \quad (8.30)$$

155

because z is an odd function under a parity transformation. In other words a system of charges which has a centre of inversion symmetry cannot have a permanent electric dipole moment.

However, an electric field can induce in an atom an electric dipole moment, proportional to E_z, and give rise to an energy of interaction proportional to E_z^2. In second order perturbation theory the energy shift of the unperturbed state $|\gamma J M_J\rangle$ is

$$\Delta W = e^2 E_z^2 \sum_{\gamma' J'} \frac{|\langle \gamma J M_J| z |\gamma' J' M_J\rangle|^2}{W_{\gamma J} - W_{\gamma' J'}}. \tag{8.31}$$

(To avoid confusion in this section we have switched to the notation W for energy.) Equation (8.31) describes the *quadratic Stark effect*. The quantum numbers represented by γ and γ' must be such that the states $|\gamma J M_J\rangle$ and $|\gamma' J' M_J\rangle$ have opposite parity, and the matrix element in eq. (8.31) is diagonal in M_J because z commutes with J_z.

Bearing in mind that, in the central-field approximation, the levels $W_{\gamma' J'}$ and $W_{\gamma J}$ belong to different configurations because of the parity requirement, we see that the perturbation procedure is often justified because the numerator in eq. (8.31) is likely to be only about $(ea_0 E_z)^2$, which is of the order of $(4 \text{ cm}^{-1})^2$ in a field of 10^7 Vm^{-1}, whereas the energy denominator is large.

To find the dependence of eq. (8.31) on M_J we need to know the matrix elements of z, diagonal in M_J. They are

$$\langle \gamma J M_J| z |\gamma' J M_J\rangle = A(\gamma\gamma' J) M_J, \tag{8.32}$$

$$\langle \gamma J M_J| z |\gamma' J - 1\ M_J\rangle = B(\gamma\gamma' J)(J^2 - M_J^2)^{1/2}, \tag{8.33}$$

$$\langle \gamma J M_J| z |\gamma' J + 1\ M_J\rangle = C(\gamma\gamma' J)((J + 1)^2 - M_J^2)^{1/2}. \tag{8.34}$$

The interaction connects the level J with a level $J' = J$ or $J \pm 1$ only. In LS coupling, for which L and S are also good quantum numbers, $L' = L$ or $L \pm 1$; and $S' = S$ since z commutes with $\mathbf{S}^2$. On substituting eqs. (8.32), (8.33), and (8.34) into eq. (8.31) we obtain three terms, each summed over all levels $\gamma' J'$ of parity opposite to that of γJ. The dependence on γ, γ', and J is complicated, but the entire M_J-dependence is of the form

$$\Delta W = R - T M_J^2 \tag{8.35}$$

where $R = R(\gamma, \gamma', J)$ and $T = T(\gamma, \gamma', J)$. Thus the degeneracy in M_J is only partly lifted, for ΔW does not depend on the sign of M_J: the electric field polarizes the charge distribution of the electrons independent of the *sense* of the precession about the z-axis. The coefficient of E_z^2 in eq. (8.31) is of course related to the *polarizability* of the atom in the state $|\gamma J M_J\rangle$.

Let us now consider as an example the quadratic Stark effect in the

8.2. Quadratic Stark effect

sodium D-lines† (see Fig. 8.7). Unlike the Zeeman effect, the Stark effect gives an unsymmetrical displacement of the lines from the position of the unperturbed line. The ground state $^2S_{1/2}$ $(M_J = \pm\frac{1}{2})$ is bound to be shifted downwards in second-order perturbation theory because of repulsion by

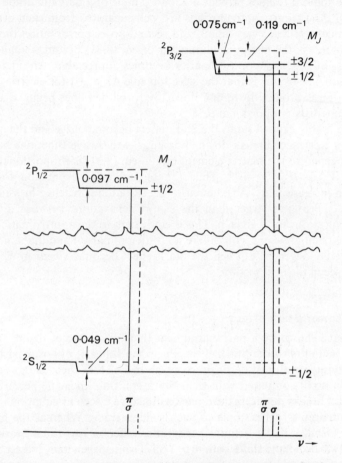

Fig. 8.7. The quadratic Stark effect in the sodium D lines. The levels and the transitions in zero electric field are indicated by dashed lines. The displacements of the levels are shown for an electric field of 250 kV/cm. These data are taken from the paper by H. Kopfermann and W. Paul, *Z. Phys.* **120**, 545, 1943.

the states above it; and since the perturbing states of opposite parity lie a long way away, giving large energy denominators in eq. (8.31), one expects the Stark shift of the ground state to be small. The states of the ^2P term are also shifted downwards by the ^2S and ^2D terms—actually more by the ^2D,

† H. Kopfermann and W. Paul, *Z. Phys.* **120**, 545, 1943, from whose paper the data are taken. The experiment was done in absorption.

the nearest of which is 12,000 cm^{-1} away, than by the ^{2}S. The selection rules for M_J are the same as in the Zeeman effect, but because of the remaining degeneracy in M_J the light emitted parallel to the electric field is unpolarized.

The sodium D-lines provide a good example of a pure quadratic Stark effect: the various configurations are well separated from each other so the Stark shifts are very small, and second-order perturbation theory is adequate. As the energy denominator of eq. (8.31) becomes smaller the mixing of states of opposite parity becomes appreciable. An interesting consequence of this is that the selection rule $\Delta l = \pm 1$ for electric dipole radiation is broken down and normally forbidden lines begin to appear with intensity proportional to E_z^2.

For highly excited states the Stark effect becomes large and the perturbation treatment breaks down. For large n the wave functions become hydrogenic, so the matrix element of z in eq. (8.31) is proportional to n^2 and $W_{\gamma J} - W_{\gamma' J'} \propto n^{-3}$. Thus the second-order expression for ΔW would increase as n^7. As n increases the situation is reached in which the Stark splitting is larger than the energy separations between different configurations for a given n. This is equivalent to the approximation that all the configurations with different l (different parity) and the same n are effectively degenerate. Then the Stark effect becomes *linear* in E_z as we shall see in the next section.

8.3. Linear Stark effect

We have shown by a parity argument that for a state of definite parity there is no first-order Stark effect. But when states of opposite parity are degenerate or almost degenerate, which means that their energy separation is small compared with their Stark splitting, then the perturbation method breaks down; instead one has to solve a secular equation.

Hydrogen is an example of such a degeneracy. Whereas the ground state $1s\ ^2S_{1/2}$ has a definite parity and suffers a quadratic Stark shift downwards, all the states with $n > 1$ have some degeneracy in l according to the Dirac theory of the fine structure. This degeneracy is removed only when the very small Lamb shift is taken into consideration. We shall assume that the electric field is sufficiently large that the entire fine structure is less than the Stark splitting: thus we have to deal with the effectively degenerate group of states of a given n (>1). (For very small fields, such that the Stark splitting is much less than the Lamb shift, states of opposite parity are not degenerate and one would expect a quadratic Stark effect. The actual situation would be complicated by hyperfine structure effects.)

The simplest case is that with $n = 2$. We can ignore electron spin, and consider the wave functions ψ_{nlm_l} of which there are four: ψ_{200}, ψ_{211},

8.3. Linear Stark effect

ψ_{210}, ψ_{21-1}. These four are degenerate eigenfunctions of the zeroth-order Hamiltonian

$$\mathscr{H}_0 = -\frac{\hbar^2}{2m}\nabla^2 - \frac{e^2}{4\pi\varepsilon_0 r} \tag{8.36}$$

as in eq. (2.30). The perturbing Hamiltonian $\mathscr{H}_E = eE_z z$ (eq. (8.29)) commutes with l_z, so all matrix elements of $\mathscr{H}_E$ vanish except for those with m_l the same, i.e., $m_l = 0$; thus there are non-vanishing matrix elements of $\mathscr{H}_E$ between the states ψ_{200} and ψ_{210}. These two wave functions also satisfy the condition that they are to be of opposite parity.

We find the perturbed energy levels in this degenerate situation by solving a secular equation for the 2×2 matrix with $m_l = 0$:

$$\begin{vmatrix} \mathscr{H}_{aa} - W & \mathscr{H}_{ab} \\ \mathscr{H}_{ba} & \mathscr{H}_{bb} - W \end{vmatrix} = 0. \tag{8.37}$$

The matrix elements are

$$\mathscr{H}_{aa} = \int \psi_{200}^* \mathscr{H}_E \psi_{200}\, d\tau = 0, \tag{8.38}$$

$$\mathscr{H}_{ab} = \mathscr{H}_{ba} = \int \psi_{200}^* \mathscr{H}_E \psi_{210}\, d\tau = -3ea_0 E_z, \tag{8.39}$$

$$\mathscr{H}_{bb} = \int \psi_{210}^* \mathscr{H}_E \psi_{210}\, d\tau = 0. \tag{8.40}$$

The result of eq. (8.39) can easily be verified by direct integration with the wave functions

$$\psi_{200} = (32\pi)^{-1/2}(a_0)^{-3/2}(2 - r/a_0)\, e^{-r/2a_0}, \tag{8.41}$$

$$\psi_{210} = (32\pi)^{-1/2}(a_0)^{-3/2}(r/a_0)\, e^{-r/2a_0} \cos\theta, \tag{8.42}$$

(see problem 8.4). Thus the secular equation (8.37) becomes

$$\begin{vmatrix} -W & -3ea_0 E_z \\ -3ea_0 E_z & -W \end{vmatrix} = 0, \tag{8.43}$$

with the solutions

$$W = \pm 3ea_0 E_z. \tag{8.44}$$

In a field of $10^7\ \mathrm{Vm}^{-1}$ the magnitude of these energy shifts is about $12\ \mathrm{cm}^{-1}$, which is large compared with the fine structure splitting, $\sim 0.35\ \mathrm{cm}^{-1}$, of the $n = 2$ level of hydrogen. Thus fairly modest fields satisfy the condition for a linear Stark effect, namely that the Stark splitting is large compared with the energy separation between states of opposite parity.

In the approximation that the Stark effect is linear the two states ψ_{211} and ψ_{21-1} are not shifted. The two states corresponding to the energy shifts $W = \pm 3ea_0E_z$ are linear combinations of the states ψ_{200} and ψ_{210}. Just as in the example of chapter 5 these combinations are found to be (see problem 8.5)

$$W = +3ea_0E_z: \psi_+ = (2)^{-1/2}(\psi_{200} - \psi_{210}); \qquad (8.45)$$

$$W = -3ca_0E_z: \psi_- = (2)^{-1/2}(\psi_{200} + \psi_{210}). \qquad (8.46)$$

The function ψ_+ corresponds to an electron charge distribution whose centre of charge is displaced by an amount $3a_0$ along the positive z-direction relative to the nucleus, and conversely for ψ_-. It is in this sense that these states of the hydrogen atom, degenerate in l, have electric dipole moments, giving rise to a linear Stark effect. The states ψ_+ and ψ_- are *not* associated with a definite orbital angular momentum about the nucleus.

8.4. Relative intensities in the Zeeman effect

In this section we want to discuss the relative intensities of the Zeeman components of a line in electric dipole radiation. We have already derived in chapter 3 the polarization rules for a hydrogen-like atom and we have stated the selection rules, eq. (8.16), for M_J in the more general case of a many-electron atom. One of the results of chapter 3 was eq. (3.56) in which the transition probability for electric dipole radiation was given for unpolarized light. But since we are now to be concerned with the direction of polarization we must refrain from averaging over all directions of polarization and go back to an earlier equation such as eq. (3.43) in which we see that for electric dipole radiation we need the matrix element squared

$$|\langle j| \, \hat{e} \cdot \mathbf{r} \, |i\rangle|^2$$

where $\hat{e}$ defines the direction of polarization. For a transverse wave, $\hat{e}$ is perpendicular to the direction of propagation of the wave, $\mathbf{k}$:

$$\hat{e} \cdot \mathbf{k} = 0. \qquad (8.47)$$

We shall discuss the electric-dipole transition from the upper state $|\gamma J M_J\rangle$ to the lower state $|\gamma' J' M_J'\rangle$ in the presence of a magnetic field $\mathbf{B}$ which defines the axis of quantization. We assume that we have a Zeeman effect linear in B, that is, in zeroth order J and J' are good quantum numbers. But we make no assumption about the sets of quantum numbers

8.4. Relative intensities in the Zeeman effect

γ and γ' other than that they describe states of opposite parity. Thus our results will have wide generality, following from the properties of the angular momentum $\mathbf{J}$ and its projection M_J. Before we discuss the quantities $|\langle \gamma'J'M'_J| \hat{\mathbf{e}} \cdot \mathbf{r} |\gamma JM_J\rangle|^2$ which determine the relative intensities of the Zeeman components let us clarify the geometrical features of the problem.

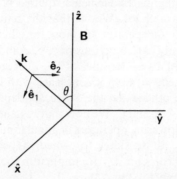

Fig. 8.8. The orthogonal set of vectors $\mathbf{k}$, $\hat{\mathbf{e}}_1$, and $\hat{\mathbf{e}}_2$ with $\mathbf{k}$ and $\hat{\mathbf{e}}_1$ lying in the x–z plane.

Let the direction of $\mathbf{B}$ be chosen as the z-axis of a Cartesian co-ordinate system labelled by unit vectors $\hat{\mathbf{x}}$, $\hat{\mathbf{y}}$, and $\hat{\mathbf{z}}$ (see Fig. 8.8), and let the wave-vector $\mathbf{k}$ make an angle θ with the z-axis. Since we have cylindrical symmetry about the z-axis we impose no loss of generality if we take $\mathbf{k}$ to lie in the x–z plane. Further, let $\hat{\mathbf{e}}$ be decomposed into $\hat{\mathbf{e}}_1$ and $\hat{\mathbf{e}}_2$:

$$\hat{\mathbf{e}} = a_1\hat{\mathbf{e}}_1 + a_2\hat{\mathbf{e}}_2 \tag{8.48}$$

where $\hat{\mathbf{e}}_1$, $\hat{\mathbf{e}}_2$, and $\mathbf{k}$ form an orthogonal set of axes with $\hat{\mathbf{e}}_1$ lying in the x–z plane. The coefficients a_1 and a_2 satisfy a normalization condition

$$|a_1|^2 + |a_2|^2 = 1 \tag{8.49}$$

and in general they contain the phases necessary to describe elliptically polarized radiation. We can also express $\hat{\mathbf{e}}$ in terms of its components along the $\hat{\mathbf{x}}$, $\hat{\mathbf{y}}$, $\hat{\mathbf{z}}$ directions:

$$\hat{\mathbf{e}} = a_1 \cos\theta\, \hat{\mathbf{x}} + a_2\hat{\mathbf{y}} - a_1 \sin\theta\, \hat{\mathbf{z}}. \tag{8.50}$$

Since $\mathbf{r} = x\hat{\mathbf{x}} + y\hat{\mathbf{y}} + z\hat{\mathbf{z}}$ we could proceed to find the relative intensities of the Zeeman components in terms of the matrix elements of x, y, and z, but we prefer not to, because the matrix elements are more conveniently written in terms of the so-called spherical tensor components of $\mathbf{r}$:

$$
\begin{aligned}
r_0 &= z, \\
r_{-1} &= (2)^{-1/2}(x - iy), \\
r_{+1} &= -(2)^{-1/2}(x + iy).
\end{aligned} \tag{8.51}
$$

In spherical tensor notation the definition of the unit vectors $\hat{e}_q (q = 0, \pm 1)$ is such that the vector $\mathbf{r} = \sum_q (-1)^q r_q \hat{e}_{-q}$. Thus with the definitions $\hat{e}_0 = \hat{z}, \hat{e}_{-1} = (2)^{-1/2}(\hat{x} - i\hat{y}), \hat{e}_{+1} = -(2)^{-1/2}(\hat{x} + i\hat{y})$, we have

$$\mathbf{r} = r_0 \hat{e}_0 - r_{-1} \hat{e}_{+1} - r_{+1} \hat{e}_{-1}$$
$$= r_0 \hat{z} + r_{-1}(2)^{-1/2}(\hat{x} + i\hat{y}) - r_{+1}(2)^{-1/2}(\hat{x} - i\hat{y}). \qquad (8.52)$$

The non-vanishing matrix elements squared of these components of $\mathbf{r}$ are written out in table 8.1. We see that the only non-vanishing matrix elements of this vector operator are those for which $J' = J$ or $J \pm 1$. The quantities A, B, and C are functions of γ, γ', J and of a radial integral. In dealing with the relative intensities of the Zeeman components of a line for which ΔJ is fixed, we are not concerned with the magnitudes of these quantities. On the other hand the entire M_J-dependence is contained in the coefficients of A, B, and C. We note that r_{+1} and r_{-1} share the property of the raising and lowering operators in angular momentum theory in that they connect only those states for which $M_J' = M_J \pm 1$ respectively. Thus the selection rules for M_J are implicit in the matrix elements. Since the Zeeman components for $\Delta M_J = 0, \pm 1$ are separated in frequency according to eq. (8.18) or, as a special case, eq. (8.17), we can associate each matrix element with a particular Zeeman component. This situation would not have been so obvious had we chosen to work with the matrix elements of x, y, and z.

Table 8.1. Matrix elements squared of the components of $\mathbf{r}$

For $J' = J$:

$$|\langle\gamma' J M_J | z | \gamma J M_J\rangle|^2 = M_J{}^2 A$$
$$|\langle\gamma' J M_J + 1 | r_{+1} | \gamma J M_J\rangle|^2 = \tfrac{1}{2}(J - M_J)(J + M_J + 1)A$$
$$|\langle\gamma' J M_J - 1 | r_{-1} | \gamma J M_J\rangle|^2 = \tfrac{1}{2}(J + M_J)(J - M_J + 1)A$$
$$\sum_{M_J} |\langle\gamma' J M_J' | \mathbf{r} | \gamma J M_J\rangle|^2 = J(J + 1)A$$

For $J' = J + 1$:

$$|\langle\gamma' J + 1 M_J | z | \gamma J M_J\rangle|^2 = \{(J + 1)^2 - M_J{}^2\}B$$
$$|\langle\gamma' J + 1 M_J + 1 | r_{+1} | \gamma J M_J\rangle|^2 = \tfrac{1}{2}(J + M_J + 1)(J + M_J + 2)B$$
$$|\langle\gamma' J + 1 M_J - 1 | r_{-1} | \gamma J M_J\rangle|^2 = \tfrac{1}{2}(J - M_J + 1)(J - M_J + 2)B$$
$$\sum_{M_J} |\langle\gamma' J + 1 M_J' | \mathbf{r} | \gamma J M_J\rangle|^2 = (J + 1)(2J + 3)B$$

For $J' = J - 1$:

$$|\langle\gamma' J - 1 M_J | z | \gamma J M_J\rangle|^2 = \{J^2 - M_J^2\}C$$
$$|\langle\gamma' J - 1 M_J + 1 | r_{+1} | \gamma J M_J\rangle|^2 = \tfrac{1}{2}(J - M_J)(J - M_J - 1)C$$
$$|\langle\gamma' J - 1 M_J - 1 | r_{-1} | \gamma J M_J\rangle|^2 = \tfrac{1}{2}(J + M_J)(J + M_J - 1)C$$
$$\sum_{M_J} |\langle\gamma' J - 1 M_J' | \mathbf{r} | \gamma J M_J\rangle|^2 = J(2J - 1)C$$

8.4. Relative intensities in the Zeeman effect

We can now write out an expression for $|\langle \gamma' J' M_J'| \, \hat{\mathbf{e}} \cdot \mathbf{r} \, |\gamma J M_J\rangle|^2$ by combining the geometrical part of this problem, eq. (8.50), with what we might call the spectroscopic part, i.e., the matrix elements of table 8.1, via eq. (8.52). We find (see problem 8.7)

$$
\begin{aligned}
|\langle \gamma' J' M_J'| \, \hat{\mathbf{e}} \cdot \mathbf{r} \, |\gamma J M_J\rangle|^2 = & \; |a_1|^2 \sin^2 \theta \, |\langle \gamma' J' M_J'| \, z \, |\gamma J M_J\rangle|^2 \\
& + \tfrac{1}{2}\{|a_1|^2 \cos^2 \theta + |a_2|^2 - i(a_1 a_2^* - a_1^* a_2) \cos \theta\} \\
& \hspace{3cm} |\langle \gamma' J' M_J'| \, r_{-1} \, |\gamma J M_J\rangle|^2 \\
& + \tfrac{1}{2}\{|a_1|^2 \cos^2 \theta + |a_2|^2 + i(a_1 a_2^* - a_1^* a_2) \cos \theta\} \\
& \hspace{3cm} |\langle \gamma' J' M_J'| \, r_{+1} \, |\gamma J M_J\rangle|^2 \quad (8.53)
\end{aligned}
$$

in which, for a given M_J, only one of the matrix elements on the right-hand side remains as soon as M_J' is given an allowed value. In this expression we have before us a description of the relative intensity in electric dipole radiation for a direction of observation given by θ, for a state of polarization given by $|a_1|$, $|a_2|$ and their phase difference, and for a Zeeman component $|\gamma J M_J\rangle \rightarrow |\gamma' J' M_J'\rangle$. In special cases eq. (8.53) can be greatly simplified: we shall consider only transverse ($\theta = \pi/2$) and longitudinal ($\theta = 0$) observation.

In transverse observation we have

$$
\begin{aligned}
|\langle \gamma' J' M_J'| \, \hat{\mathbf{e}} \cdot \mathbf{r} \, |\gamma J M_J\rangle|^2 = & \; |a_1|^2 |\langle \gamma' J' M_J'| \, z \, |\gamma J M_J\rangle|^2 \\
& + \tfrac{1}{2}|a_2|^2 |\langle \gamma' J' M_J'| \, r_{-1} \, |\gamma J M_J\rangle|^2 \\
& + \tfrac{1}{2}|a_2|^2 |\langle \gamma' J' M_J'| \, r_{+1} \, |\gamma J M_J\rangle|^2
\end{aligned}
$$

$$\text{transverse} \quad (8.54)$$

and in longitudinal observation

$$
\begin{aligned}
|\langle \gamma' J' M_J'| \, \hat{\mathbf{e}} \cdot \mathbf{r} \, |\gamma J M_J\rangle|^2 & \\
= & \; \tfrac{1}{2}\{1 - i(a_1 a_2^* - a_1^* a_2)\}|\langle \gamma' J' M_J'| \, r_{-1} \, |\gamma J M_J\rangle|^2 \\
& + \tfrac{1}{2}\{1 + i(a_1 a_2^* - a_1^* a_2)\}|\langle \gamma' J' M_J'| \, r_{+1} \, |\gamma J M_J\rangle|^2
\end{aligned}
$$

$$\text{longitudinal} \quad (8.55)$$

In eq. (8.55) the normalization condition (8.49) has been used. These two formulae (8.54) and (8.55) can be applied at once to any particular example (see problem 8.8).

Let us discuss these two cases in further detail. In transverse observation the polarization condition $a_2 = 0$, $a_1 = 1$, which corresponds to the electric vector parallel to the z-axis (π polarization), is associated with the selection rule $\Delta M_J = 0$; and the condition $a_1 = 0$, $a_2 = 1$, which corresponds to the electric vector in the x–y plane, is associated with the selection rule $\Delta M_J = \pm 1$. Thus we see that the (geometrical) polarization rules are explicitly connected with the (spectroscopic) selection rules.

163

The relative intensities are:

$$\pi \quad \Delta M_J = 0 \quad |\langle \gamma' J' M_J | z | \gamma J M_J \rangle|^2,$$
$$\Delta M_J = -1 \quad \tfrac{1}{2} |\langle \gamma' J' M_J - 1 | r_{-1} | \gamma J M_J \rangle|^2, \quad (8.56)$$
$$\sigma \quad \Delta M_J = +1 \quad \tfrac{1}{2} |\langle \gamma' J' M_J + 1 | r_{+1} | \gamma J M_J \rangle|^2.$$

Notice that for $\Delta M_J = \pm 1$ there is a factor $\tfrac{1}{2}$ in the geometrical part and, from table 8.1, another factor $\tfrac{1}{2}$ in the matrix element squared. In longitudinal observation one can distinguish between the transitions $\Delta M_J = +1$ and $\Delta M_J = -1$ as regards their polarization as follows: let $a_1 = \alpha$ and $a_2 = (1 - \alpha^2)^{1/2} e^{i\delta}$ where δ is a phase difference. Then

$$i(a_1 a_2^* - a_1^* a_2) = 2\alpha(1 - \alpha^2)^{1/2} \sin \delta \quad (8.57)$$

and the geometrical factors in eq. (8.55) become $\tfrac{1}{2}(1 \mp 2\alpha(1 - \alpha^2)^{1/2} \sin \delta)$. With $\alpha = (2)^{-1/2}$, the contributions of r_{-1} and r_{+1} vanish respectively when $\delta = \pi/2$ or $-\pi/2$. In these two cases each non-vanishing component is identified with circularly polarized radiation of opposite sense (σ_- and σ_+):

$$\delta = \pi/2: \quad |\langle \gamma' J' M_J' | \hat{\mathbf{e}} \cdot \mathbf{r} | \gamma J M_J \rangle|^2$$
$$= I^l_{\sigma_-} = |\langle \gamma' J' M_J - 1 | r_{-1} | \gamma J M_J \rangle|^2;$$
$$\delta = -\pi/2: \quad |\langle \gamma' J' M_J' | \hat{\mathbf{e}} \cdot \mathbf{r} | \gamma J M_J \rangle|^2$$
$$= I^l_{\sigma_+} = |\langle \gamma' J' M_J + 1 | r_{+1} | \gamma J M_J \rangle|^2. \quad (8.58)$$

(We have used the notation I^l to denote relative intensities in longitudinal observation.) In contrast with eq. (8.56) there is no factor $\tfrac{1}{2}$ in the geometrical part of eq. (8.58).

Let us now discuss certain sum rules. We have written out in table 8.1 the result for a given line, i.e., γJ, $\gamma' J'$ fixed:

$$\sum_{M_J} |\langle \gamma' J' M_J' | \mathbf{r} | \gamma J M_J \rangle|^2 \text{ is independent of } M_J. \quad (8.59)$$

The notation here is

$$|\langle j | \mathbf{r} | i \rangle|^2 = |\langle j | z | i \rangle|^2 + |\langle j | r_+ | i \rangle|^2 + |\langle j | r_- | i \rangle|^2.$$

Equation (8.59) is the important result already quoted in chapter 3, eq. (3.59). It states that the rate of decay, and hence the lifetime, of each M_J-state belonging to a level γJ is the same. We assume, of course, that the frequencies of the Zeeman components are so nearly equal that the ω^3 dependence of the transition rate is the same for all. This assumption we take to be implicit in the use of first-order perturbation theory.

Another result, which can be verified from table 8.1, is that in transverse observation

$$\sum_{M_J} \sum_{M_J'} I^t_\pi = \sum_{M_J} \sum_{M_J'} I^t_\sigma, \quad (8.60)$$

8.4. Relative intensities in the Zeeman effect

where

$$I_\pi^t = |\langle \gamma' J' M'_J| \, z \, |\gamma J M_J\rangle|^2 \qquad (8.61)$$

and

$$I_\sigma^t = \tfrac{1}{2} |\langle \gamma' J' M'_J| \, r_+ \, |\gamma J M_J\rangle|^2 + \tfrac{1}{2} |\langle \gamma' J' M'_J| \, r_- \, |\gamma J M_J\rangle|^2. \qquad (8.62)$$

We have used here the notation I^t to denote relative intensities in transverse observation. Equation (8.60) states the requirement that in the limit of zero field (or in the limit that the Zeeman components are unresolved spectroscopically) the radiation is unpolarized. Indeed, one must expect that an analogous result is true for any angle of observation θ. From eq. (8.53) one can verify (see problem 8.9) that

$$\sum_{M_J} \sum_{M'_J} I(\theta; a_1 = 1, a_2 = 0) = \sum_{M_J} \sum_{M'_J} I_\pi^t, \qquad (8.63)$$

and

$$\sum_{M_J} \sum_{M'_J} I(\theta; a_1 = 0, a_2 = 1) = \sum_{M_J} \sum_{M'_J} I_\sigma^t, \qquad (8.64)$$

both independent of θ. With eq. (8.60) these equations lead in the limit to unpolarized radiation for any angle θ. In longitudinal observation we have

$$\sum_{M_J} \sum_{M'_J} I_{\sigma+}^l = \sum_{M_J} \sum_{M'_J} I_{\sigma-}^l = \sum_{M_J} \sum_{M'_J} I_\sigma^l. \qquad (8.65)$$

The fundamental eq. (8.59) can be re-written in terms of the relative intensities in transverse observation, I_π^t and I_σ^t. With eqs. (8.61) and (8.62), eq. (8.59) becomes

$$\sum_{M_J} \{I_\pi^t + 2I_\sigma^t\} \text{ is independent of } M_J. \qquad (8.66)$$

Equations (8.60) and (8.65) combined read:

$$\sum_{M_J} \sum_{M'_J} I_\pi^t = \sum_{M_J} \sum_{M'_J} I_\sigma^t = \sum_{M_J} \sum_{M'_J} I_{\sigma+}^l = \sum_{M_J} \sum_{M'_J} I_{\sigma-}^l. \qquad (8.67)$$

These two sum rules, eqs. (8.66) and (8.67), together with the assumption that the Zeeman intensity pattern is symmetrical, are sufficient to give the relative intensities of Zeeman components. Indeed, they were used originally† with the correspondence principle to derive the formulae of table 8.1. The use of the sum rules is neat and convenient for finding relative intensities in transitions involving small values of J (see problem 8.10), but for larger J it is much more straightforward to evaluate the matrix elements.

It is helpful to bear in mind the properties of a radiating classical dipole in this discussion of relative intensities, for otherwise the various factors

† H. Hönl, *Z. Phys.* **31**, 340, 1925.

of $\frac{1}{2}$, which we have pointed out as they appeared, might be confusing. For example, in eq. (8.66) we have to give double weight to I_σ^t relative to I_π^t. Now we have formulated the problem as if we had a classical dipole with a π-component oscillating in the z-direction and two σ-components rotating in opposite senses in the x–y plane. One of these rotating components can be resolved into two oscillating components, each of relative amplitude $(2)^{-1/2}$, one in the x-direction and one in the y-direction. If, as we have assumed, the x-direction is the direction of transverse observation only the y-oscillator contributes to the observed I_σ^t, with relative intensity $\frac{1}{2}$. But it is important to realize that eq. (8.66) refers to what is emitted, not merely to what is observed in the x-direction. Therefore, since $\sum_{M_j} I_\sigma^t$ represents only half what is emitted with σ-polarization we must restore full weight with a factor 2. Similar considerations relate the absence of a factor $\frac{1}{2}$ in eq. (8.58) to its presence in eq. (8.62).

Finally, the generality of the theory of angular momentum allows us to apply the formulae of table 8.1 to many circumstances. The representation $|\gamma J M_J\rangle$ may describe the Zeeman states of a fine-structure level under any coupling scheme provided the field is weak. We could equally well discuss the case when L and S are completely decoupled by writing the state as $|\gamma L M_L\rangle$. We have incorporated S in γ because in electric dipole radiation $\Delta S = 0$, $\Delta M_S = 0$, and formally we replace J, M_J by L, M_L in the formulae. In particular, the sum rules show at once that for singlet spectra ($S = 0$) in transverse observation the π-component is twice as strong as each σ component. The formulae of table 8.1 also apply to one-electron states $|l m_l\rangle$, though of course parity considerations forbid the transition $l \rightarrow l$.

Problems

(Those problems marked with an asterisk are more advanced.)

8.1. Estimate the magnitude of a 'strong' field in the Zeeman effect of the $n = 2$ fine structure levels in hydrogen.

8.2. Evaluate the g_J-factors for the levels of the $5p6s$ configuration in j–j coupling. There is a g-sum rule which states that for a given configuration the sum of the g-values of all levels with the same J is independent of the coupling scheme. Confirm this rule for the $5p6s$ configuration in LS and j–j coupling.

*8.3. What is the form of the vector potential $\mathbf{A}$ which represents a uniform magnetic field $\mathbf{B} = B\mathbf{k}$ (in the z-direction), where $\mathbf{B} = \text{curl } \mathbf{A}$ and $\text{div } \mathbf{A} = 0$?

For such a field show that replacement of the operator $\mathbf{p}^2/2m$ by $(1/2m)(\mathbf{p} + e\mathbf{A})^2$ in the Hamiltonian for a one-electron atom leads to a description of (a) the Zeeman effect for orbital motion and (b) diamagnetism. [The electron charge is $-e$.]

Problems

8.4. Use the Schrödinger hydrogenic functions ψ_{nlm_l} to show that for $n = 2$ in hydrogen

$$\int \psi^*_{200} eE_z z \psi_{210} \, d\tau = -3ea_0 E_z,$$

which is eq. (8.39). Thus the Stark operator, $eE_z z$, has off-diagonal matrix elements in the nlm_l representation between degenerate states of opposite parity.

*8.5. For $n = 2$, $m_l = 0$ in hydrogen there are two functions, ψ_+ and ψ_-, which diagonalize the Stark operator $eE_z z$. Find these functions as linear combinations of ψ_{200} and ψ_{210}, where $\psi_\pm$ correspond to the Stark energy shifts $W = \pm 3ea_0 E_z$ respectively: see eqs. (8.45) and (8.46). In the absence of an electric field ψ_+ and ψ_- are still orthogonal solutions of Schrödinger's equation of definite energy, but they are not associated with definite orbital angular momentum.

Sketch graphs of ψ_+ and ψ_- as a function of z. Hence show qualitatively that hydrogen in the state ψ_+ can be said to have a negative electric dipole moment induced by an applied electric field, but one which leads to an energy shift linear in electric field.

8.6. For hydrogen in an electric field the effective potential energy is

$$V = -e^2/4\pi\varepsilon_0 r + eE_z z.$$

Sketch V as a function of z and discuss the effects of tunnelling through a potential barrier. (See H. Kuhn, *Atomic Spectra*, pp.109 ff.)

*8.7. Verify eq. (8.53). Remember that r_+ and r_- are not Hermitian operators and that $\langle i| \, r_- \, |j \rangle^* = \langle j| \, r^\dagger_- \, |i \rangle$ and that the Hermitian adjoint of r_-, written $r^\dagger_-$, is $r^\dagger_- = -r_+$.

8.8. Use eqs. (8.54) and (8.55), or more particularly eqs. (8.56) and (8.58), together with table 8.1 to evaluate the relative intensities of the Zeeman components of the sodium D lines in transverse and in longitudinal observation.

8.9. Verify eqs. (8.63) and (8.64).

8.10. Use the sum rules, eqs. (8.66) and (8.67), to evaluate the relative intensities of the Zeeman components of the sodium D lines in transverse and in longitudinal observation.

8.11. Derive classically expressions for the frequencies of a Lorentz triplet in the normal Zeeman effect. To do this, consider the modification, brought about by the application of a magnetic field, in the motion of a system of three-dimensional harmonic oscillators. Notice that (a) the charge to mass ratio of an oscillator must be that of an electron if the result is to agree with experiment; and (b) the mechanical frequency of the oscillator, which is equated to the frequency of electric dipole radiation, is involved in the calculation—the derivation by Lorentz in 1897 predated the concept of quantized energy levels in atoms.

9. Hyperfine structure and isotope shift

So far we have considered the nucleus only as a massive point charge, responsible for the large electrostatic interaction with the charged electrons which surround it and for the smaller electrodynamic interaction with the spinning electrons *moving* through the electric field of the nuclear charge (i.e., spin–orbit interaction).

There is a further splitting of energy levels, called hyperfine structure, which is extremely small. The range of splittings is of the order of 10^{-3}–1 cm^{-1}, or 30–30,000 Mc/s. The lower end of this range is at the limit of resolution of optical spectroscopic methods in which the line-width is governed predominantly by Doppler broadening, but it lies conveniently within the scope of modern radiofrequency methods of spectroscopy. The technical problem of measuring optically the difference in frequency between two electric dipole transitions at, say 20,000 cm^{-1} when the frequency difference, 10^{-3}–1 cm^{-1}, is of the order of the line width is certainly a formidable one. On the other hand the direct observation of magnetic dipole transitions between hyperfine structure levels is what is achieved in radiofrequency spectroscopy: the precision is limited by the natural line-width, which can be made very small for ground states of free atoms.

We explain hyperfine structure in terms of properties of the nucleus other than its charge. In 1924 Pauli suggested that a nucleus has a total angular momentum, which is labelled by the quantum number I. This quantum number may have integral or half-integral values, like the total electronic angular momentum quantum number J, for a nucleus is a compound structure of nucleons—protons and neutrons—each of which has an intrinsic spin $\frac{1}{2}$ and may take part in orbital motion within the nucleus. We shall consider the total nuclear angular momentum quantum number I, or 'nuclear spin', as fixed for a given isotope because we shall only treat one energy state of the nucleus, namely the ground state. We attribute to a nucleus, in addition to its spin, electromagnetic multipole moments of higher order than electric monopole. The interaction between these moments and the electromagnetic field produced at the nucleus by the electrons is responsible for hyperfine structure.

By symmetry arguments of parity and time-reversal the only multipole (2^k-pole) moments which do not vanish are: the magnetic moments for odd k, and the electric moments for even k, i.e. (apart from electric charge, $k = 0$), magnetic dipole ($k = 1$), electric quadrupole ($k = 2$), magnetic octupole ($k = 3$), electric hexadecapole ($k = 4$), etc. The approximate spherical symmetry of an atomic system makes this multipole expansion a convenient one: the angular part of such an expansion is expressed in terms of spherical harmonics of order k. Interaction of the higher order multipole moments ($k \geqslant 3$) with the electrons is interesting but often negligibly small: in fact only a few electric hexadecapole interactions have been observed. We shall confine ourselves to the two lowest orders of interaction: that of a nuclear magnetic dipole moment with an electronic magnetic field, and that of a nuclear electric quadrupole moment with an electronic gradient of electric field. These two interactions turn out to be of the same order of magnitude for non-s electrons, as we shall see, and they are about 10^8 times larger than the octupole and hexadecapole interactions.

9.1. Magnetic dipole interaction

We consider first the interaction of a point nuclear magnetic moment $\mathbf{\mu}_I$ with the magnetic field $\mathbf{B}_{el}$ produced by the electrons at the nucleus. We write a Hamiltonian

$$\mathcal{H} = -\mathbf{\mu}_I \cdot \mathbf{B}_{el} \qquad (9.1)$$

which is to be treated as a small perturbation. The assumption in eq. (9.1) is that $\mathbf{\mu}_I$ depends on nuclear co-ordinates only and that $\mathbf{B}_{el}$ depends on electronic co-ordinates only. The dot product between the nuclear vector $\mathbf{\mu}_I$ and the electronic vector $\mathbf{B}_{el}$ is of course characteristic of a dipole term in a multipole expansion. As an approximation we shall assume throughout this chapter that the zeroth order Hamiltonian contains the central field, the electrostatic repulsion terms between electrons, and the spin–orbit interaction, so that we have to deal with separate electronic energy levels labelled by J. That is to say we assume that we are dealing with an isolated level J, or in particular that the hyperfine structure is small compared with the fine structure. We shall not consider the interesting exceptions to this case which occur particularly in heavy elements. Thus we can call our approximation the IJ coupling approximation (in analogy with LS coupling): I and J are good quantum numbers, and we confine ourselves to considering matrix elements of the perturbation, eq. (9.1), which are diagonal in I and J.

An immediate consequence of the IJ coupling approximation is that

169

we can write the vector operator $\boldsymbol{\mu}_I$ as proportional to $\mathbf{I}$:

$$\boldsymbol{\mu}_I = \frac{\boldsymbol{\mu}_I \cdot \mathbf{I}}{I(I + 1)} \mathbf{I}, \tag{9.2}$$

just as in eq. (8.10) for $\boldsymbol{\mu}_J$. Through this equation we may define an effective g-factor for the nucleus:

$$\boldsymbol{\mu}_I = g_I \mu_N \mathbf{I}. \tag{9.3}$$

According to the convention which we are adopting this equation differs in sign from the analogous eq. (8.11) for g_J. g_I is positive if $\boldsymbol{\mu}_I$ lies along $\mathbf{I}$. Also the unit μ_N for nuclear magnetic moments is the nuclear magneton

$$\mu_N = \frac{e\hbar}{2M} = \frac{m_0}{M} \mu_B = \frac{\mu_B}{1{,}836 \cdot 13} \tag{9.4}$$

where M is the proton mass. Thus in these units g_I is a number of the order of unity. Sometimes eq. (9.3) is written

$$\boldsymbol{\mu}_I = g_I' \mu_B \mathbf{I} \tag{9.3a}$$

in units of Bohr magnetons in which case g_I' is a very small number. It is perhaps less confusing to avoid the use of g_I altogether and to write simply

$$\boldsymbol{\mu}_I = (\mu_I / I)\mathbf{I} \tag{9.3b}$$

where μ_I is the value of the nuclear moment, quoted usually in nuclear magnetons.

It also follows immediately from the adoption of the *IJ* coupling approximation that we can write $\mathbf{B}_{el} \propto \mathbf{J}$ since we have already assumed that the vector operator $\mathbf{B}_{el}$ operates in the space of the electronic co-ordinate only. Thus eq. (9.1) is of the form

$$\mathscr{H} = A\mathbf{I} \cdot \mathbf{J} \tag{9.5}$$

where A is a parameter to be determined from experiment. We shall return to this equation later, but first we shall discuss the detailed form of $\mathbf{B}_{el}$ for a single-electron atom.

We consider a single-electron atom for which $l \neq 0$. Then semi-classically the magnetic field at the nucleus consists of two parts: that due to the orbital motion of the electron of charge $-e$, co-ordinate $\mathbf{r}$, about the nucleus as origin, and that due to the spin magnetism of the electron at a distance r from the nucleus:

$$\mathbf{B}_{el} = \frac{\mu_0}{4\pi} \frac{(-e\mathbf{v}) \times (-\mathbf{r})}{r^3} - \frac{\mu_0}{4\pi} \frac{1}{r^3} \left[\boldsymbol{\mu}_s - \frac{3(\boldsymbol{\mu}_s \cdot \mathbf{r})\mathbf{r}}{r^2} \right], \quad r \neq 0, \tag{9.6}$$

where $-\mathbf{r}$ is the co-ordinate of the nucleus with respect to the electron.

9.1. Magnetic dipole interaction

With $\boldsymbol{\mu}_s = -2\mu_B \mathbf{s}$ (with $g_s = 2$ exactly) and $-e\mathbf{r} \times \mathbf{v} = -2\mu_B \mathbf{l}$,

$$\mathbf{B}_{el} = -2\frac{\mu_0}{4\pi}\frac{\mu_B}{r^3}\left[\mathbf{l} - \mathbf{s} + \frac{3(\mathbf{s}\cdot\mathbf{r})\mathbf{r}}{r^2}\right], \quad l \neq 0, \tag{9.7}$$

and the perturbing Hamiltonian of eq. (9.1) is

$$\mathcal{H} = \left(\frac{\mu_0}{4\pi}\right)\left(2\mu_B\frac{\mu_I}{I}\right)\frac{\mathbf{I}\cdot\mathbf{N}}{r^3}, \tag{9.8}$$

where

$$\mathbf{N} = \mathbf{l} - \mathbf{s} + 3(\mathbf{s}\cdot\mathbf{r})\mathbf{r}/r^2. \tag{9.9}$$

The first-order energy shift arising from this perturbation is the expectation value of $\mathcal{H}$. The zeroth-order wave functions, separable in nuclear and electronic co-ordinates, are $|\gamma I M_I j m_j\rangle$ which are $(2I + 1)(2j + 1)$-fold degenerate in M_I and m_j. As usual for degenerate perturbation theory we form new zeroth-order functions $|\gamma I j F M_F\rangle$ which are linear combinations of the functions $|\gamma I M_I j m_j\rangle$ where

$$\mathbf{F} = \mathbf{I} + \mathbf{j}. \tag{9.10}$$

The quantum number F describes the total angular momentum of the atom. F remains a good quantum number under the application of the perturbation $\mathcal{H}$, for the hyperfine interaction does not give a torque on the atom as a whole. Therefore

$$\Delta E = \langle \gamma I j F M_F | \mathcal{H} |\gamma I j F M_F\rangle. \tag{9.11}$$

To evaluate ΔE we project $\mathbf{N}$ on to $\mathbf{j}$ since we are taking matrix elements diagonal in j, and $\mathcal{H}$ becomes effectively

$$\mathcal{H} = \left(\frac{\mu_0}{4\pi}\right)\left(2\mu_B\frac{\mu_I}{I}\right)\frac{\mathbf{N}\cdot\mathbf{j}}{j(j+1)}\frac{\mathbf{I}\cdot\mathbf{j}}{r^3}. \tag{9.12}$$

Then

$$\Delta E = \left(\frac{\mu_0}{4\pi}\right)\left(2\mu_B\frac{\mu_I}{I}\right)\left\langle\frac{\mathbf{N}\cdot\mathbf{j}}{j(j+1)r^3}\right\rangle\tfrac{1}{2}\{F(F+1) - j(j+1) - I(I+1)\} \tag{9.13}$$

$$= \frac{a_j}{2}\{F(F+1) - j(j+1) - I(I+1)\} \tag{9.14}$$

where

$$a_j = \left(\frac{\mu_0}{4\pi}\right)\left(2\mu_B\frac{\mu_I}{I}\right)\left\langle\frac{\mathbf{N}\cdot\mathbf{j}}{j(j+1)r^3}\right\rangle, \quad l \neq 0. \tag{9.15}$$

The angular form of eq. (9.14) is quite analogous to that of eq. (7.52) for fine structure, and we shall discuss it shortly. Meanwhile let us investigate a_j which fixes the magnitude of the hyperfine structure splitting.

From eq. (9.9.) $\mathbf{N} \cdot \mathbf{j}$, which is the same as $\mathbf{N} \cdot (\mathbf{l} + \mathbf{s})$, is

$$\mathbf{N} \cdot \mathbf{j} = (\mathbf{l} - \mathbf{s} + 3(\mathbf{s} \cdot \mathbf{r})\mathbf{r}/r^2) \cdot (\mathbf{l} + \mathbf{s})$$
$$= \mathbf{l}^2 - \mathbf{s}^2 + 3(\mathbf{s} \cdot \mathbf{r})(\mathbf{r} \cdot \mathbf{l})/r^2 + 3(\mathbf{s} \cdot \mathbf{r})(\mathbf{r} \cdot \mathbf{s})/r^2. \quad (9.16)$$

Now $\mathbf{r} \cdot \mathbf{l} = 0$ since $\hbar\mathbf{l} = \mathbf{r} \times \mathbf{p}$, so the third term in eq. (9.16) vanishes. Also

$$\mathbf{s}^2 - 3(\mathbf{s} \cdot \mathbf{r})^2/r^2 = 0, \quad (9.17)$$

which can be shown by writing out components and making use of the commutation relations for spin, or alternatively by invoking the relation (see problem 4.2(c))

$$(\mathbf{s} \cdot \mathbf{F})(\mathbf{s} \cdot \mathbf{G}) = \tfrac{1}{4}\mathbf{F} \cdot \mathbf{G} + \tfrac{1}{2}(i/\hbar)\mathbf{s} \cdot (\mathbf{F} \times \mathbf{G}) \quad (9.18)$$

where $\mathbf{F}$ and $\mathbf{G}$ are any two vector operators which commute with $\mathbf{s}$ (here we put $\mathbf{F} = \mathbf{r}$ and $\mathbf{G} = \mathbf{r}$). Finally, therefore, eq. (9.16) becomes

$$\mathbf{N} \cdot \mathbf{j} = \mathbf{l}^2 \quad (9.19)$$

and

$$a_j = \left(\frac{\mu_0}{4\pi}\right)\left(2\mu_\mathrm{B}\frac{\mu_I}{I}\right)\langle 1/r^3 \rangle \frac{l(l+1)}{j(j+1)}, \quad l \neq 0. \quad (9.20)$$

To complete the picture we must consider the interaction between the nuclear magnetic dipole moment and an s-electron ($l = 0$) for which it turns out, though this is not immediately obvious, that a_j of eq. (9.20) vanishes. There is an interaction between the nuclear moment and the intrinsic spin magnetic moment of an s-electron which depends upon the fact that an s-electron has a non-zero probability density at the origin $|\psi(0)|^2$. This is called the Fermi contact interaction, and it has the form

$$\mathcal{H} = a_s\mathbf{I} \cdot \mathbf{s}$$
$$= a_s\mathbf{I} \cdot \mathbf{j} \quad (9.21)$$

since $\mathbf{j} = \mathbf{s}$ for $l = 0$, where

$$a_s = \left(\frac{\mu_0}{4\pi}\right)\left(2\mu_\mathrm{B}\frac{\mu_I}{I}\right)(8\pi/3)|\psi(0)|^2, \quad l = 0. \quad (9.22)$$

One might think that a full relativistic treatment is necessary for the derivation of this formula. However, a demonstration, if not a rigorous derivation, of the right answer is given by the following semi-classical considerations: for an s-electron there is a spherically symmetrical distribution of spin magnetism which does not vanish at the origin. The magnetization, or spin magnetic moment per unit volume, at the origin is

$$\mathbf{P}_0 = \boldsymbol{\mu}_s|\psi(0)|^2. \quad (9.23)$$

9.1. Magnetic dipole interaction

Since the magnetic field at the origin arising from a spherical *shell* of uniform magnetization vanishes, the resulting field is the same as that due to a sphere of uniform magnetization $\mathbf{P}_0$,

$$\mathbf{B} = (2\mu_0/3)\mathbf{P}_0 = (2\mu_0/3)\boldsymbol{\mu}_s|\psi(0)|^2 \qquad (9.24)$$

independent of the radial variation of $\psi^2(r)$ provided there is spherical symmetry (see problem 9.1). The magnetic interaction of a point nuclear magnetic moment with this field is

$$\mathcal{H} = -(2\mu_0/3)\boldsymbol{\mu}_I \cdot \boldsymbol{\mu}_s|\psi(0)|^2 \qquad (9.25)$$

which gives the expression (9.22) for a_s. It is interesting to regard this demonstration as a confirmation that $\mathbf{B} = (2\mu_0/3)\mathbf{P}_0$ and not $\mathbf{H} = -(1/3)\mathbf{P}_0$ is to be used in describing this atomic phenomenon, for otherwise there would be disagreement with experiment. In other words, $\mathbf{B}$, which is associated classically with solenoidal fields arising from circulating currents, is appropriate here.

For hydrogen we can evaluate a_s and a_j. For $l = 0$

$$|\psi(0)|^2 = \frac{Z^3}{\pi a_0^3 n^3}, \qquad (9.26)$$

hence

$$a_s = \left(\frac{\mu_0}{4\pi}\right)\left(2\mu_B \frac{\mu_I}{I}\right)\frac{8}{3}\frac{Z^3}{a_0^3 n^3}, \quad l = 0. \qquad (9.27)$$

For $l \neq 0$

$$\langle 1/r^3 \rangle = \frac{Z^3}{a_0^3 n^3 (l + \frac{1}{2})l(l + 1)}, \qquad (9.28)$$

hence

$$a_j = \left(\frac{\mu_0}{4\pi}\right)\left(2\mu_B \frac{\mu_I}{I}\right)\frac{Z^3}{a_0^3 n^3}\frac{1}{(l + \frac{1}{2})j(j + 1)}, \quad l \neq 0. \qquad (9.29)$$

(It turns out that a_j gives a_s for $l = 0$, $j = \frac{1}{2}$ but this is another result special to hydrogen.)

The hyperfine structure of the $1s\ ^2S_{1/2}$ ground level of hydrogen is interesting because it is calculable and has been measured to high precision by the method of magnetic resonance in an atomic beam. The nuclear spin (proton spin) is $I = \frac{1}{2}$. Therefore from eq. (9.14) the ground level splits into two levels $F = 1$ and $F = 0$ separated by an energy interval $h\Delta v = a_s$. Substitution of the proton magnetic moment $\mu_I = 2.79275$ n.m. (nuclear magnetons) in eq. (9.27) for a_s, together with a reduced mass correction, leads to a value of Δv differing by about 1 part in 1,000, much larger than experimental error, from the experimental result. There are other more refined corrections to be made, but historically the most important is the introduction by Breit of the anomalous magnetic moment of the electron. His suggestion that $g_s \neq 2$ stimulated work on quantum electrodynamics to give the result

$$g_s = 2\left(1 + \frac{\alpha}{2\pi} - 0{\cdot}328 \frac{\alpha^2}{\pi^2} \cdots\right)$$
$$= 2 \times 1{\cdot}0011596. \qquad (9.30)$$

This correction in particular greatly improves the agreement with the experimental result† which, with the development of the hydrogen maser, has surpassed all other frequency measurements in precision:

$$\Delta v\,(\text{H}) = 1{,}420,\ 405,\ 751{\cdot}7662 \pm 0{\cdot}0030\ \text{Hz}$$

referred to the frequency standard of a caesium atomic beam clock

$$\Delta v\,(\text{Cs}) = 9{,}192,\ 631,\ 770\ \text{Hz}.$$

Before we consider the hyperfine structure for other simple configurations let us write down the form of the magnetic dipole interaction for many electrons. In the non-relativistic approximation we can write, summing the $\mathbf{B}_{\text{el}}$ of eqs. (9.7) and (9.24) over all electrons,

$$\mathscr{H} = \left(\frac{\mu_0}{4\pi}\right)\left(2\mu_{\text{B}}\frac{\mu_I}{I}\right)\mathbf{I}\cdot\sum_i\left\{\frac{\mathbf{l}_i}{r_i^3} - \frac{1}{r_i^3}\left(\mathbf{s}_i - \frac{3(\mathbf{s}_i\cdot\mathbf{r}_i)\mathbf{r}_i}{r_i^2}\right)\right.$$
$$\left. + (8\pi/3)|\psi_i(0)|^2\mathbf{s}_i\right\}, \qquad (9.31)$$

where, because of the spherical symmetry of the closed shells, we need only consider a summation over valence electrons. In the absence of configuration mixing and relativistic effects (which are very important in hyperfine structure, but which we shall not be able to discuss) the first two terms in eq. (9.31), the orbital and spin-dipole terms, apply for electrons with $l \neq 0$, and the third term, the contact term, applies for electrons with $l = 0$. In any case, for a level with well-defined total electronic angular momentum J, eq. (9.31) can be written

$$= A(J)\mathbf{I}\cdot\mathbf{J} \qquad (9.32)$$

in which the single parameter $A(J)$ is proportional to μ_I/I and to B_{el}, the magnitude of the magnetic field at the nucleus produced by the electrons.

9.2. Determination of nuclear spin from magnetic hyperfine structure

To first order the energy shift of a level J is given by the expectation value of the Hamiltonian of eq. (9.32):

$$\Delta E = \langle\gamma IJFM_F|\,A(J)\mathbf{I}\cdot\mathbf{J}\,|\gamma IJFM_F\rangle$$
$$= \tfrac{1}{2}A(J)\{F(F+1) - J(J+1) - I(I+1)\}, \qquad (9.33)$$

† L. Essen, R. W. Donaldson, E. G. Hope, and M. J. Bangham, *Metrologia*, **9**, 128, 1973. See also S. B. Crampton, D. Kleppner, and N. F. Ramsey, *Phys. Rev. Letters* **11**, 338, 1963. Earlier work had been conducted by P. Kusch: see *Phys. Rev.* **100**, 1188, 1955.

9.2. Determination of nuclear spin

which reduces to eq. (9.14) for a single electron. As already remarked, the analogy with spin–orbit interaction giving rise to fine structure is exact with the replacements

$$J \to F$$
$$L \to J$$
$$S \to I$$
$$\zeta(^{2S+1}L) \to A(J).$$

(a) There is a splitting into hyperfine structure levels, labelled by F, which are $(2F + 1)$-fold degencrate;

(b) $A(J)$ is a measure of the splitting and there is an *interval rule*:

$$\Delta E(F) - \Delta E(F - 1) = A(J)F; \qquad (9.34)$$

(c) there is a sum rule for relative intensities;

(d) F has $2I + 1$ values for $I < J$ or $2J + 1$ values for $J \leqslant I$;

(e) the electric dipole selection rule for F is $\Delta F = 0, \pm 1, F = 0 \nrightarrow F = 0$.

Let us consider, as an example of the measurement of nuclear spin from optical hyperfine structure, the work of Anderson† on ^{139}La. He studied

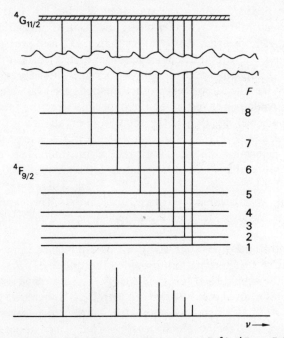
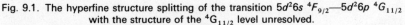

Fig. 9.1. The hyperfine structure splitting of the transition $5d^2 6s \, ^4F_{9/2}$—$5d^2 6p \, ^4G_{11/2}$ with the structure of the $^4G_{11/2}$ level unresolved.

† O. E. Anderson, *Phys. Rev.* **45**, 685, 1934.

the hyperfine splitting of the line $5d^26s$ $^4F_{9/2}$—$5d^26p$ $^4G_{11/2}$ at 6,250 Å with a Fabry–Perot interferometer, and deduced that $I(^{139}La) = \frac{7}{2}$. The interpretation of the spectrum is made simpler if one can find a transition in which one of the levels has a large hyperfine structure and the other has negligible structure. The magnetic dipole interaction is largest for configurations with an unpaired s-electron which gives rise to a contact interaction (see problem 9.2 for a confirmation of this in the formulae for hydrogen). Thus, neglecting the contribution of the $5d^2$ electrons we assume that the $5d^26s$ configuration has a reasonably large structure and the $5d^26p$ has only a very small (unresolved) structure. The structure of the spectral line therefore reflects the structure of the level which is split (see Fig. 9.1). The nuclear spin I can be found by counting the number of components of the line if $J > I$, for then the number of components is simply $2I + 1$. Thus the transition under consideration is very suitable for determining I because $J = \frac{9}{2}$ which is reasonably large, so that if there are less than $2J + 1 = 10$ components, I is less than J. Anderson was able to count at least seven resolved components, but the eighth component in the high-frequency tail of the pattern was unresolved so the counting method gave only $I > \frac{5}{2}$ definitely.

Anderson also studied the relative intensities of the components, which should be proportional to the statistical weights, $2F + 1$, of the final levels. He had difficulty in measuring the relative intensities accurately enough to establish that $I = \frac{7}{2}$ rather than $I = \frac{9}{2}$ (see problem 9.3). The difficulty in achieving sufficient accuracy in intensity measurements is not untypical of optical spectroscopic work, and intensity measurements should not be relied upon when other methods are available.

Finally Anderson investigated the hyperfine structure intervals. From the interval rule, eq. (9.34), the relative intervals are proportional to $F = I + J, I + J - 1, \ldots, |J - I| + 1$, so if J is known the intervals lead to a determination of I. With the largest interval set equal to 100 arbitrary units, the six resolved relative intervals were 100, $86\cdot8 \pm 1\cdot3$, $74\cdot4 \pm 1\cdot4$, $61\cdot2 \pm 1\cdot7$, $51\cdot7 \pm 1\cdot8$, $38\cdot5 \pm 2\cdot1$. These intervals are consistent only with a value $I = \frac{7}{2}$ (see problem 9.3). This experiment illustrates three ways of determining a nuclear spin from optical measurements.

Notice that the *sign* of $A(^4F_{9/2})$ can be established immediately from the hyperfine structure pattern provided the split level $^4F_{9/2}$ is known to lie below the unsplit level $^4G_{11/2}$ in energy. The hyperfine structure of the level $^4F_{9/2}$ is normal (not inverted) and $A(^4F_{9/2})$ is therefore positive. The value of $A(^4F_{9/2})$ is in fact about $0\cdot014$ cm^{-1}, the largest interval is $0\cdot111$ cm^{-1}, and the whole structure spreads over about $0\cdot500$ cm^{-1}. These numbers are typical for hyperfine structures in moderately heavy elements.

176

9.3. Determination of μ_I from magnetic hyperfine structure

In general the magnetic dipole interaction constant $A(J)$ is proportional to μ_I/I and B_{el}, so to find μ_I (with I known) one has to calculate B_{el} which is related to $1/r^3$ for each electron. A direct calculation of $\langle 1/r^3 \rangle$ with radial wave functions is not at all reliable—one cannot trust the wave functions to give answers to better than ± 25 per cent except in some modern work on special cases. Some of the most accurate values of nuclear moments have been obtained directly by nuclear magnetic resonance, independent of electronic behaviour to a first approximation, and indeed some recent work on hyperfine structure has used these accurate values of μ_I to evaluate the electronic parameter $\langle 1/r^3 \rangle$ and hence to throw light on the accuracy of radial calculations.

In discussing the determination of μ_I/I from $A(J)$ we shall confine ourselves to the alkali atoms. Goudsmit and Fermi and Segré have modified the hydrogenic formulae for a_j and a_s to take account of the screened central field in alkalis. For $l \neq 0$, a_j of eq. (9.20) contains the parameter $\langle 1/r^3 \rangle$. The *fine-structure* doublet separation ΔW, which contains $\langle 1/r \cdot dV/dr \rangle$ can be expressed in the form (eq. 7.74))

$$\Delta W = \left(\frac{\mu_0}{4\pi}\right) 2\mu_B^2 Z_i \langle 1/r^3 \rangle (l + \tfrac{1}{2}) \tag{9.35}$$

where $Z_i e$ is the effective charge introduced by Landé, and provided this fine structure parameter $\langle 1/r^3 \rangle$ can be regarded as the same as the hyperfine structure parameter, we have from eq. (9.20)

$$a_j = \left(\frac{\mu_I}{\mu_B I}\right) \frac{l(l + 1)}{j(j + 1)(l + \tfrac{1}{2})} \frac{\Delta W}{Z_i}. \quad l \neq 0. \tag{9.36}$$

In terms of Landé's formula (7.75) for ΔW, we have

$$a_j = \left(\frac{\mu_I}{\mu_B I}\right) \frac{\alpha^2 R_\infty Z_i Z_0^2}{n^{*3} j(j + 1)(l + \tfrac{1}{2})}, \quad l \neq 0. \tag{9.37}$$

For s-electrons similarly we use a modified $|\psi(0)|^2$:

$$|\psi(0)|^2 = \frac{Z_i Z_0^2}{\pi a_0^3 n^{*3}} \tag{9.38}$$

with which eq. (9.22) gives

$$a_s = \left(\frac{\mu_0}{4\pi}\right)\left(2\mu_B \frac{\mu_I}{I}\right) \frac{8}{3} \frac{Z_i Z_0^2}{a_0^3 n^{*3}}, \tag{9.39}$$

or in different units

$$a_s = \left(\frac{\mu_I}{\mu_B I}\right) \frac{8}{3} \alpha^2 R_\infty \frac{Z_i Z_0^2}{n^{*3}}. \tag{9.40}$$

The empirical values $Z_i = Z$ for s-electrons, $Z_i = Z - 4$ for p-electrons, and $Z_i = Z - 11$ for d-electrons have been found to be appropriate in formulas (9.37) and (9.40), but these non-relativistic expressions do not, as they stand, match the experimental results. Relativistic corrections are important particularly: (a) for s- and $p_{1/2}$-electrons because, in a description of their motion close to the nucleus, the non-relativistic approximation that the kinetic energy is much less than mc^2 breaks down seriously; (b) for large Z for which, in a central field, the relativistic factor $(1 - v^2/c^2)^{1/2}$ is related to $(1 - \alpha^2 Z^2)^{1/2}$. A summary of relativistic and other corrections can be found in Kopfermann's† book.

To show the orders of magnitude of the corrections we compare in the table below Na I ($Z = 11$), Cs I ($Z = 55$) and the alkali-like Bi V ($Z = 83$). In the second column of the table is given the ratio $a_s(\text{relativistic})/a_s(\text{non-relativistic})$, and in the third column the ratio $(a_{p1/2}/a_{p3/2})$ (relativistic) which, from eq. (9.37), has the value 5 in the non-relativistic approximation.

	$\dfrac{a_s(\text{rel})}{a_s(\text{non-rel})}$	$(a_{p1/2}/a_{p3/2})(\text{rel})$
Na I ($Z = 11$)	1·04	$5 \times 1·01$
Cs I ($Z = 55$)	1·5	$5 \times 1·3$
Bi V ($Z = 83$)	2·3	$5 \times 2·0$

Thus, for s- and $p_{1/2}$-electrons and large Z, factors of two or more are introduced. The corrected formulae for alkali-like spectra have led to remarkably accurate evaluations of nuclear moments from measurements of the magnetic hyperfine structure.

We have not discussed all the refinements in the problem of relating μ_I/I to A. But we conclude this section by mentioning one anomalous effect, the so-called Bohr–Weisskopf‡ effect. We might expect that for two isotopes the ratio of the A factors is equal to the ratio of the μ_I/I. This is not so, but rather $A_1/A_2 = (1 + \Delta)(\mu_I/I)_1/(\mu_I/I)_2$ where Δ is called the hyperfine structure anomaly for the two isotopes 1 and 2. This effect is one result of the finite size of the nucleus: the nuclear magnetic dipole moment must be regarded as distributed over the volume of the nucleus, so the interaction $-\mu_I \cdot \mathbf{B}_{el}$ must be averaged over the nuclear volume. The differential effect between isotopes of different nuclear size can be significant for electrons which penetrate the nucleus (s- and $p_{1/2}$-electrons), leading to anomalies of the order of $\Delta \sim 1$ per cent. Obviously the effect is only revealed when really accurate measurements of μ_I and of A are made independently. Nevertheless, a considerable amount of work on the hyperfine structure anomaly has been done and has given further information about nuclear structure.

† H. Kopfermann, *Nuclear Moments*, Academic Press, 1958.
‡ A. Bohr and V. F. Weisskopf, *Phys. Rev.* 77, 94, 1950.

9.4. Magnetic hyperfine structure in two-electron spectra

In first-order perturbation theory the energy shift arising from the magnetic dipole interaction in two-electron spectra is just the expectation value of the perturbing Hamiltonian, eq. (9.31). In the IJ coupling approximation the zeroth-order wave functions are labelled by I and J as good quantum numbers. The experimentally determined parameter $A(J)$ can be related to the a-factors for the individual electrons provided an appropriate coupling scheme for the electrons is chosen.

We shall treat one simple example, namely the $nln's$ configuration, in the LS coupling approximation. The reason for not attempting a more general many-electron problem in this book is that the matrix elements of the second term in eq. (9.31), the spin-dipole part, are not easily evaluated by elementary methods. The configuration $nln's$ is an important one to discuss in any case, because the unpaired s-electron gives rise to an appreciable hyperfine structure in comparison with which the contribution of the nl electron can often be neglected. We adopt this approximation and put

$$a_{nl} \approx 0.$$

If we give the nl electron the suffix 1 and the $n's$ electron the suffix 2 we can write the combined wave function as $|L(s_1s_2)SJ; I; FM_F\rangle$ where $\mathbf{L} = \mathbf{l}_1$, $\mathbf{l}_2 = 0$, $\mathbf{s}_1$ and $\mathbf{s}_2$ are coupled up to their resultant $\mathbf{S}$, $\mathbf{L}$ and $\mathbf{S}$ have the resultant $\mathbf{J}$, and $\mathbf{J}$ and $\mathbf{I}$ have the resultant $\mathbf{F}$. The Hamiltonian of eq. (9.31) has only one term

$$\mathcal{H} = a_{n's}\mathbf{I} \cdot \mathbf{s}_2, \tag{9.41}$$

where

$$a_{n's} = \left(\frac{\mu_0}{4\pi}\right)\left(2\mu_B \frac{\mu_I}{I}\right)\frac{8\pi}{3}|\psi_{n's}(0)|^2. \tag{9.42}$$

In taking matrix element diagonal in S and J, we project $\mathbf{s}_2$ on $\mathbf{S}$ first (LS coupling) and then $\mathbf{S}$ on $\mathbf{J}$ (IJ coupling) to obtain

$$\Delta E = \left\langle L(s_1s_2)SJ; I; FM_F \left| a_{n's}\mathbf{I} \cdot \left(\frac{(\mathbf{s}_2 \cdot \mathbf{S})(\mathbf{S} \cdot \mathbf{J})}{S(S + 1)J(J + 1)}\mathbf{J}\right)\right| \right.$$
$$\times L(s_1s_2)SJ; I; FM_F\rangle$$
$$= a_{n's}\frac{1}{2}\left\{\frac{S(S + 1) + s_2(s_2 + 1) - s_1(s_1 + 1)}{S(S + 1)}\right\}$$
$$\times \frac{1}{2}\left\{\frac{J(J + 1) + S(S + 1) - L(L + 1)}{J(J + 1)}\right\}$$
$$\times \langle JIFM_F| \mathbf{I} \cdot \mathbf{J} |JIFM_F\rangle. \tag{9.43}$$

On the other hand, the Hamiltonian is also

$$\mathcal{H} = A(J)\mathbf{I} \cdot \mathbf{J} \tag{9.44}$$

in the IJ coupling approximation, so

$$A(J) = a_{n's} \frac{J(J + 1) + S(S + 1) - L(L + 1)}{4J(J + 1)} \tag{9.45}$$

since $s_2(s_2 + 1)$ and $s_1(s_1 + 1)$ both have the value $\frac{3}{4}$. Thus we have related $A(J)$ to $a_{n's}$.

The terms belonging to the configuration $nln's$ are ^3L and ^1L. Remembering that $S = 0$ and $J = L$ for the singlet we have

$$A(^1L_J) = 0 \tag{9.46}$$

and

$$A(^3L_J) = a_{n's} \frac{J(J + 1) + 2 - L(L + 1)}{4J(J + 1)},$$
$$J = L + 1, L, \text{ or } L - 1. \tag{9.47}$$

The vanishing of $A(^1L_J)$ is a consequence of the way in which $\mathbf{s}_1$ is coupled to $\mathbf{s}_2$ to give $S = 0$ so that the magnetic field at the nucleus averages to zero. Thus for example the line $5s5p\ ^1P_1$—$5s5d\ ^1D_2$ at 6,438 Å in the spectrum of Cd I is a very narrow line because even under high resolution the hyperfine structure arising from the $5p$ electron (and still less from the $5d$ electron) is too small to be resolved. In fact this line was once used as a wavelength standard. (There is also, in principle, an unresolved structure due to isotope shift in natural cadmium which consists of several isotopes but this structure is very small.)

9.5. Electric quadrupole interaction

The electrostatic interaction between a proton, charge e, with co-ordinate $\mathbf{r}_n$, and an electron, charge $-e$, with co-ordinate $\mathbf{r}_e$ is

$$\mathcal{H} = \frac{-e^2}{4\pi\varepsilon_0|\mathbf{r}_e - \mathbf{r}_n|} \tag{9.48}$$

where the origin of co-ordinates is the centre of mass of the (infinitely heavy) nucleus. If we sum over the protons of the nucleus and over the electrons we have the total electrostatic interaction between the nucleus and the electrons.

So far we have considered the nucleus as a point charge ($r_n \ll r_e$) with spherical symmetry. In order to be able to discuss departures of the nuclear charge distribution from spherical symmetry we should like to attribute to the nucleus electric multipole moments. To do this we have to separate the nuclear from the electronic co-ordinates. We can achieve a

9.5. Electric quadrupole interaction

separation if we assume $r_e > r_n$, that is if we ignore for the time being penetration of the nuclear volume by the electrons. With this assumption eq. (9.48) can be expanded in powers of r_n/r_e and in Legendre polynomials $P_k(\cos\theta_{en})$ where θ_{en} is the angle between $\mathbf{r}_e$ and $\mathbf{r}_n$:

$$\mathcal{H} = -\frac{e^2}{4\pi\varepsilon_0}\{r_e^2 + r_n^2 - 2r_e r_n \cos\theta_{en}\}^{-1/2}$$

$$= -\frac{e^2}{4\pi\varepsilon_0}\sum_k \frac{r_n^k}{r_e^{k+1}} P_k(\cos\theta_{en})$$

$$= -\frac{e^2}{4\pi\varepsilon_0 r_e} - \frac{e^2 r_n}{4\pi\varepsilon_0 r_e^2} P_1(\cos\theta_{en}) - \frac{e^2 r_n^2}{4\pi\varepsilon_0 r_e^3} P_2(\cos\theta_{en}) \ldots \quad (9.49)$$

The first term represents a monopole interaction which, when summed over the protons and electrons, becomes $-\sum_i Ze^2/4\pi\varepsilon_0 r_{e_i}$. We have already discussed this spherically symmetric term as the nuclear part of the central field interaction, and we shall now omit it. The higher order terms represent electric dipole, electric quadrupole interactions, etc., and we assume that the series converges sufficiently rapidly that we need not consider terms of order higher than $k = 2$. We can complete the separation of co-ordinates by applying the spherical harmonic addition theorem

$$P_k(\cos\theta_{en}) = \frac{4\pi}{2k+1}\sum_{q=-k}^{k}(-1)^q Y_k^{-q}(\theta_n, \phi_n) Y_k^q(\theta_e, \phi_e) \quad (9.50)$$

where the Y_k^q are spherical harmonics of rank k, projection q. These harmonics have been given in eq. (2.42): the integers k and q are formally analogous to l and m_l respectively in eq. (2.42), and the spherical polar

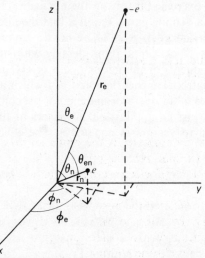

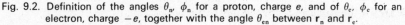

Fig. 9.2. Definition of the angles θ_n, ϕ_n for a proton, charge e, and of θ_e, ϕ_e for an electron, charge $-e$, together with the angle θ_{en} between $\mathbf{r}_n$ and $\mathbf{r}_e$.

co-ordinates θ and ϕ are defined with respect to an arbitrary z-axis fixed in space (see Fig. 9.2). The electric quadrupole term of eq. (9.49) can now be written

$$\sum_{q=-2}^{2} (-1)^q \left\{ \left(\frac{4\pi}{5}\right)^{1/2} er_n^2 Y_2^{-q}(n) \right\} \left\{ \left(\frac{4\pi}{5}\right)^{1/2} \left(\frac{-e}{4\pi\varepsilon_0 r_e^3}\right) Y_2^q(e) \right\}$$
$$\equiv \sum_q (-1)^q Q_2^{-q}(n) F_2^q(e) \quad (9.51)$$

and when summed over the protons and electrons this is the nuclear electric quadrupole interaction with the electrons. It is the scalar product of a nuclear and an electronic tensor, each of rank two, and as it stands is rather complicated.

Matters are greatly simplified if we assume, as usual, that the IJ coupling approximation is valid. This implies that the nucleus has a definite nuclear spin I and the direction of $\mathbf{I}$ establishes an axis of cylindrical symmetry. In this angular momentum representation we only deal with matrix elements diagonal in I, and we can take the direction of $\mathbf{I}$ to be the principal axis for the quadrupole tensor. Now we can define a nuclear electric quadrupole *moment* in terms of the expectation value, in the state $|I, M_I = I\rangle$, of the nuclear part $Q_2^0(n)$ of the operator of eq. (9.51). From table 2.1, $(4\pi/5)^{1/2} Y_2^0(\theta_n, \phi_n) = \frac{1}{2}(3\cos^2\theta_n - 1)$, and the conventional definition of the nuclear electric quadrupole moment Q is

$$Q \equiv \frac{2}{e}\left\langle I, M_I = I \left| \sum_i Q_2^0(n)_i \right| I, M_I = I \right\rangle$$
$$= \left\langle II \left| \sum_i r_{n_i}^2 (3\cos^2\theta_{n_i} - 1) \right| II \right\rangle \quad (9.52)$$

where the sum is over the protons in the nucleus. (The assumption that the state $|II\rangle$ has a definite parity requires that the nuclear electric dipole moment, similarly defined, is zero. Therefore we can omit the second term in eq. (9.49).) The whole point about dealing with states of definite I is that all five components of the tensor operator $Q_2^{-q}(n)$ are proportional to Q and have matrix elements diagonal in I which are proportional to tensor components of an operator made up of the components of $\mathbf{I}$ itself. This statement is just an application to tensors of the general theorem which we applied to vectors in eq. (7.42). The effective operators for $Q_2^{-q}(n)$ are†

$$Q_2^0 = \frac{eQ}{2I(2I-1)}[3I_z^2 - I(I+1)],$$

$$Q_2^{\pm 1} = \mp \frac{1}{2}(6)^{1/2} \frac{eQ}{2I(2I-1)}[I_z I_\pm + I_\pm I_z], \quad (9.53)$$

$$Q_2^{\pm 2} = \frac{1}{2}(6)^{1/2} \frac{eQ}{2I(2I-1)} I_\pm^2.$$

† See N. F. Ramsey, *Molecular Beams*, O.U.P., 1956, p. 61.

9.5. Electric quadrupole interaction

In exactly the same way the average gradient of the electric field produced by the electrons at the nucleus is defined by

$$\frac{1}{e} \langle JJ| \frac{\partial^2 V_e}{\partial z^2} |JJ\rangle = \frac{2}{e} \langle JJ| \sum_i F_2^0(e)_i |JJ\rangle$$

$$= - \langle JJ| \sum_i \frac{(3\cos^2\theta_{e_i} - 1)}{4\pi\varepsilon_0 r_{e_i}^3} |JJ\rangle \qquad (9.54)$$

where the sum is over the electrons.

The product, eq. (9.51), gives the Hamiltonian for the quadrupole interaction in terms of angular momentum operators:

$$\mathscr{H}_Q = eQ \left\langle \frac{\partial^2 V_e}{\partial z^2} \right\rangle \frac{[3(\mathbf{I} \cdot \mathbf{J})^2 + \frac{3}{2}\mathbf{I} \cdot \mathbf{J} - I(I+1)J(J+1)]}{2I(2I-1)J(2J-1)} \qquad (9.55)$$

which is the quantum-mechanical form of the classical expression

$$W = \frac{1}{4} eQ \frac{\partial^2 V_e}{\partial z^2} \tfrac{1}{2}(3\cos^2\theta_{IJ} - 1) \qquad (9.56)$$

for the interaction of a nuclear electric quadrupole moment with the gradient of the electric field produced by the electrons. In eq. (9.56) θ_{IJ} is the angle between the directions of $\mathbf{I}$ and $\mathbf{J}$ which are the principal axes of the nuclear and electronic systems. The derivation of eq. (9.55) is much more complicated than that of the magnetic dipole interaction, but it is important to realize that the quadrupole interaction involves tensors of rank two.

The quadrupole moment Q is positive if the nuclear charge distribution is elongated along the direction of $\mathbf{I}$ (prolate) and negative if the distribution is flattened (oblate). The magnitude of Q is determined by $\langle r_n^2 \rangle$. As defined, it has the dimensions of area, and the unit 10^{-28} m^2 is called a 'barn'. The magnitude of $\partial^2 V_e/\partial z^2$ is determined by $\langle 1/r_e^3 \rangle$, and what is measured experimentally is the product

$$B = eQ \left\langle \frac{\partial^2 V_e}{\partial z^2} \right\rangle. \qquad (9.57)$$

Since there have been no direct measurements of the interaction of Q with a gradient of a static electric field produced in the laboratory, knowledge of Q from spectroscopic measurements of B is no more accurate than the knowledge of $\langle 1/r_e^3 \rangle$. In view of certain important corrections which have to be applied to $\langle 1/r_e^3 \rangle$, the values of nuclear electric quadrupole moments are not known at all well, say ± 25 per cent except in special cases.

The quadrupole interaction causes a shift of the hyperfine structure levels. In first order this energy shift is

$$\Delta E_Q = \langle IJFM_F| \, \mathcal{H}_Q \, |IJFM_F\rangle$$

$$= \frac{B \frac{3}{2} K(K+1) - 2I(I+1)J(J+1)}{4 \quad I(2I-1)J(2J-1)}, \qquad (9.58)$$

where

$$K = F(F+1) - J(J+1) - I(I+1). \qquad (9.59)$$

This interaction vanishes for S terms, because $\langle \partial^2 V_e/\partial z^2 \rangle$ vanishes when the electron charge distribution is spherically symmetrical. As remarked before, the interaction also vanishes unless $I \geqslant 1, J \geqslant 1$. In other terms we have, combining the magnetic dipole and electric quadrupole shifts, eqs. (9.33) and (9.58):

$$\Delta E = \frac{A}{2} K + \frac{B \frac{3}{2} K(K+1) - 2I(I+1)J(J+1)}{4 \quad I(2I-1)J(2J-1)}, \qquad (9.60)$$

which shows that the quadrupole interaction gives rise to a departure from the interval rule because its dependence on F is different from that of the magnetic dipole interaction. When B/A is sufficiently large even the order of the F-levels may be different from that when $B/A = 0$.

Let us consider the line 4,031 Å in ^{55}Mn as an example of a small departure from the interval rule. Walther† has made precise optical measurements on this line and from the results has evaluated $Q(^{55}\text{Mn})$.

Table 9.1

Interval $F - (F-1)$	Experimental interval mK	$A(^6P_{7/2})F$ mK	Departure from interval rule, mK
6—5	86·65 ± 0·11	85·74	+ 0·91
5—4	71·55 ± 0·20	71·45	+ 0·10
4—3	56·53 ± 0·20	57·16	− 0·63
3—2	41·96 ± 0·30	42·87	− 0·91

The transition is $3d^5 4s^2 \, {}^6S_{5/2}$—$3d^5 4s 4p \, {}^6P_{7/2}$. For ^{55}Mn, $I = \frac{5}{2}$. To a first approximation the S-term has no quadrupole interaction because it is spherically symmetrical, and indeed the magnetic dipole interaction vanishes also because of the symmetry of the half-filled shell d^5. (Actually $A(^6S_{5/2}) = -2\cdot415694$ mK and $B(^6S_{5/2}) = -6\cdot1 \times 10^{-4}$ mK as measured accurately by magnetic resonance in an atomic beam, but these values only lead to small corrections in the optical work.) The $^6P_{7/2}$ level, on the other hand, has both magnetic dipole and electric quadrupole hyperfine structure. The experimental intervals between F-levels, with small corrections, are given in the second column of table 9.1, and with $A(^6P_{7/2}) =$

† H. Walther, *Z. Phys.* **170**, 507, 1962.

9.5. Electric quadrupole interaction

$14 \cdot 29 \pm 0 \cdot 05$ mK the expected intervals obeying the interval rule are given in the third column. The departures from the interval rule are shown in column four, and these are a measure of $B(^6P_{7/2})$.

From these data the value $B(^6P_{7/2}) = 2 \cdot 2 \pm 0 \cdot 4$ mK is obtained. The energy level scheme is shown in Fig. 9.3, and a beautiful recording of the line profile is shown in Fig. 9.4. The value of $Q(^{55}\text{Mn})$ itself is actually about 0·35 barns, which is fairly typical as to order of magnitude.

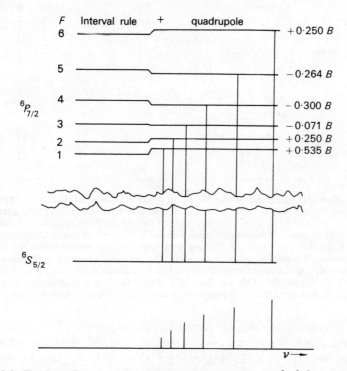

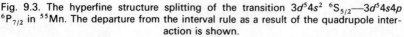

Fig. 9.3. The hyperfine structure splitting of the transition $3d^5 4s^2$ $^6S_{5/2}$—$3d^5 4s 4p$ $^6P_{7/2}$ in ^{55}Mn. The departure from the interval rule as a result of the quadrupole interaction is shown.

For a single electron we have, non-relativistically,

$$b_j = -\frac{2j - 1}{2j + 2} \frac{e^2}{4\pi\varepsilon_0} Q \left\langle \frac{1}{r_e^3} \right\rangle \tag{9.61}$$

which is the analogue for the quadrupole interaction of

$$a_j = \frac{l(l + 1)}{j(j + 1)} \left(\frac{\mu_0}{4\pi} \right) \left(2\mu_B \frac{\mu_I}{I} \right) \left\langle \frac{1}{r^3} \right\rangle, \tag{9.62}$$

for the dipole interaction, eq. (9.20). When the nuclear magnetic moment μ_I can be found directly the value of $\langle 1/r^3 \rangle$ derived from a measured

185

value of a_j in eq. (9.62) is commonly used in eq. (9.61) to evaluate Q from b_j. However, considerable care has to be taken in working out the higher order corrections to these $1/r^3$ parameters and it is the uncertainty in these corrections which effectively limits the accuracy to which values of Q can be quoted.

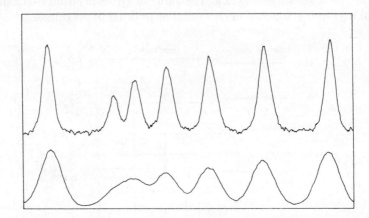

Fig. 9.4. Part of a recording of the Mn I line 4,031 Å. Range of dispersion: 416·73 mK. Upper trace: measurement with an atomic beam in absorption. The quantity $(I_0 - I)/I_0$ is recorded (I_0 is the intensity without the atomic beam and I is the intensity with the atomic beam). Lower trace: measurement with a hollow-cathode light source (current 5 mA, 0·5 Torr of Neon). The signal current from the multiplier was recorded. The wavenumber increases towards the right. The wavenumber scale is the same for both traces. By permission from Walther, H., Das Kernquadrupolmoment des ^{55}Mn, in *Zeitschrift für Physik* Bd. **170**, pp. 507–525, Berlin-Göttingen-Heidelberg: Springer 1962. Notice particularly the difference in resolution between the two traces.

It is interesting to compare the orders of magnitude of b_j and a_j for non-s electrons. We have $b_j \sim e^2 r_n^2/4\pi\varepsilon_0 a_0^3 = (r_n^2/a_0^2)e^2/4\pi\varepsilon_0 a_0$ whereas $a_j \sim (\mu_0/4\pi)\mu_B\mu_I/a_0^3 \sim (\mu_0/4\pi)(m/M)\mu_B^2/a_0^3 \sim (m/M\alpha^2\ e^2/4\pi\varepsilon_0 a_0$. Hence

$$\frac{b_j}{a_j} \sim \frac{r_n^2}{a_0^2}\left(\frac{m}{M}\alpha^2\right)^{-1} \tag{9.63}$$
$$\sim 1,$$

since $r_n^2/a_0^2 \sim 10^{-8}$. This result is quite coincidental because the magnitude of r_n^2 is determined by the strength of nuclear forces whereas a_0^2 and α^2 depend on the strength of electromagnetic forces. The 8-pole and 16-pole interactions are less than the 2-pole and 4-pole interactions by a further factor of order r_n^2/a_0^2.

9.6. Zeeman effect of hyperfine structure

The Zeeman effect of hyperfine structure in the IJ coupling approximation is concerned with the interaction of the total electronic magnetic moment $\boldsymbol{\mu}_J$ and of the nuclear magnetic moment $\boldsymbol{\mu}_I$ with a steady magnetic field along the z-axis. This interaction removes the $(2I + 1)(2J + 1)$-fold spatial degeneracy of the hyperfine structure energy levels. The formalism of this interaction is very similar to that of the Zeeman effect of fine structure, and we shall be able to quote some results by analogy, without derivation. The significant difference between the Zeeman effects in hyperfine structure and in fine structure lies in the orders of magnitude involved. Whereas we found that the strong-field case is rare in fine structure, this is not so in hyperfine structure; furthermore, $\mu_I \ll \mu_J$ and we can often neglect the direct interaction between the nuclear magnetic moment and a weak external field, whereas in the analogous situation we could not neglect μ_S in comparison with μ_L. The Zeeman effect finds an application in the determination of nuclear spins and moments by magnetic resonance methods of spectroscopy.

Let us now consider the IJ coupling approximation in which the central field and all electrostatic and magnetic interactions internal to the electron system are included in a zeroth-order Hamiltonian. We apply as a perturbation the nuclear magnetic dipole interaction and the Zeeman terms. We omit the nuclear electric quadrupole interaction to simplify the problem, *not* because this interaction is small. The reason we include the magnetic dipole interaction in the perturbation and not in the zeroth-order Hamiltonian is that it is not necessarily large compared with the Zeeman terms. The perturbation, then, is

$$
\begin{aligned}
\mathscr{H} &= A\mathbf{I} \cdot \mathbf{J} - \boldsymbol{\mu}_J \cdot \mathbf{B} - \boldsymbol{\mu}_I \cdot \mathbf{B}, \\
&= A\mathbf{I} \cdot \mathbf{J} + g_J\mu_\mathrm{B}J_zB - g'_I\mu_\mathrm{B}I_zB,
\end{aligned} \tag{9.64}
$$

where the equations

$$
\boldsymbol{\mu}_J = -g_J\mu_\mathrm{B}\mathbf{J}, \tag{9.65}
$$

$$
\boldsymbol{\mu}_I = g'_I\mu_\mathrm{B}\mathbf{I} \tag{9.66}
$$

now define the g-factors. We are using the notation in which g'_I is a small number of order $\mu_\mathrm{N}/\mu_\mathrm{B} = m/M$ (see eq. (9.3a)). The field is weak if $g_J\mu_\mathrm{B}B \ll A$, and strong if $g_J\mu_\mathrm{B}B \gg A$. If A/h is typically of the order of 1,000 MHz in frequency units, then the field is strong provided $B \gg A/\mu_\mathrm{B} = 1,000/1\cdot4 \times 10^4 \approx 0\cdot07$ T, quite a modest field.

For a weak field we apply first the perturbation $A\mathbf{I} \cdot \mathbf{J}$ and find that the $|IJFM_F\rangle$ representation is appropriate. The levels are split according to eq. (9.33). The hyperfine structure levels (I, J, F, M_F) are degenerate with respect to M_F. We now apply the Zeeman terms $g_J\mu_\mathrm{B}J_zB - g'_I\mu_\mathrm{B}I_zB$ as a perturbation and in doing so we use the representation $|IJFM_F\rangle$, for in spite of the degeneracy we can proceed as in non-degenerate perturbation theory because J_z and I_z separately commute with F_z. Hence, just as in the case of fine structure, we obtain an energy shift (see eq. (8.9) and (8.13)).

$$\Delta E = g_F \mu_B B M_F \qquad (9.67)$$

where g_F is an effective g-value analogous to eq. (8.14):

$$
\begin{aligned}
g_F = g_J &\frac{F(F + 1) + J(J + 1) - I(I + 1)}{2F(F + 1)} \\
&- g'_I \frac{F(F + 1) - J(J + 1) + I(I + 1)}{2F(F + 1)}. \qquad (9.68)
\end{aligned}
$$

Since $g'_I \ll g_J$ we can usually neglect the second term in eq. (9.68), that is we neglect the direct interaction of the nuclear magnetic moment with the *laboratory* field. On the other hand the first term in eq. (9.68) does not reduce to g_J under this approximation but contains I on an equal footing with J. This is because the interaction between the nuclear magnetic moment and the field of the *electrons* is large ($A \gg g_J \mu_B B$) in the weak field case, and in considering the interaction $-\boldsymbol{\mu}_J \cdot \mathbf{B}$ we have to project $\boldsymbol{\mu}_J$ on to $\mathbf{F}$ first. The projections involved are illustrated in the vector model picture (Fig. 9.5). The total energy shift in first order is

$$\Delta E_{F, M_F} = \tfrac{1}{2} A K + g_F \mu_B B M_F. \qquad (9.69)$$

The energy levels depend linearly on B, and eq. (9.69) is illustrated in Fig. 9.6 for the case $J = \tfrac{3}{2}, I = 3$. There are four possible values of F, and for each of these the values of g_F/g_J are

F	$\tfrac{9}{2}$	$\tfrac{7}{2}$	$\tfrac{5}{2}$	$\tfrac{3}{2}$
g_F/g_J	$\dfrac{35}{105}$	$\dfrac{25}{105}$	$\dfrac{3}{105}$	$-\dfrac{63}{105}$.

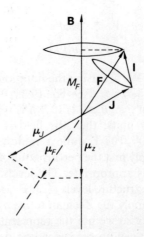

Fig. 9.5. Vector model for $\boldsymbol{\mu}$, showing the projections first on the direction of $\mathbf{F}$ and then on the z-axis. The interaction between $\boldsymbol{\mu}_I$ and the external field is neglected.

9.6. Zeeman effect of hyperfine structure

In this particular example the value of $g_{F=3/2}$ turns out to be negative. This happens when, in the classical vector triangle $\mathbf{F} = \mathbf{I} + \mathbf{J}$, the angle between $\mathbf{J}$ and $\mathbf{F}$ is obtuse. In general, for $I > J$ g_F is largest for the largest F, for $I = J$ $g_F = \frac{1}{2} g_J$, independent of F, and for $I < J$ g_F is smallest for the largest F.

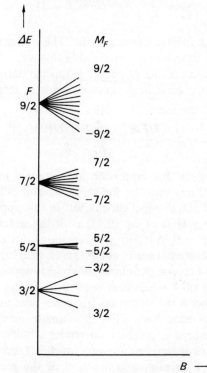

Fig. 9.6. The Zeeman effect of hyperfine structure in low magnetic field where the energy depends linearly on field. $J = \frac{3}{2}$, $I = 3$. Note the inversion of the M_F states for $F = \frac{3}{2}$.

From eq. (9.69) there follows a straightforward method of measuring I when J and g_J are known, *without a knowledge of A*. If a magnetic dipole transition, $\Delta F = 0$, $\Delta M_F = \pm 1$, can be observed, the frequency is

$$h\nu = g_F \mu_B B. \tag{9.70}$$

In particular, for $F = I + J$, $g_F = g_J J/(I + J)$ from eq. (9.68) in which g'_I is neglected, and

$$h\nu (F = I + J) = g_J \frac{J}{I + J} \mu_B B. \tag{9.71}$$

Such a transition is always observable in a magnetic resonance experiment by the method of atomic beams.† Apart from I, the only unknown in eq.

† See, for example, as a general reference N. F. Ramsey, *Molecular Beams*, O.U.P., 1956.

(9.71) is B which is usually measured in terms of another element for which g_J, J, and I are known. In particular it is sometimes possible to compare the frequencies v_1 and v_2 for two isotopes of the same element in a fixed field B, in which case

$$\frac{v_1}{v_2} = \frac{I_2 + J}{I_1 + J}, \quad (F = I + J). \tag{9.72}$$

In eq. (9.72) g_J and B have cancelled out. The nuclear spins of many radioactive isotopes have been determined in this way.

Within the limitation of the IJ coupling approximation the perturbation $g_J\mu_B B J_z$ (with the term in g_I' ignored) leads in second order to an energy shift

$$\Delta E = \sum_{F'} \frac{|\langle IJFM_F| g_J\mu_B B J_z |IJF'M_F\rangle|^2}{E_F - E_{F'}}. \tag{9.73}$$

The matrix elements in this expression are diagonal in M_F because J_z commutes with F_z. Thus states of the same M_F and different F within a manifold of given I and J 'repel each other' in this approximation, and the additional energy shift of eq. (9.73) is of the order of $(g_J\mu_B B)^2/A$. Measurement of $\Delta F = 0$, $\Delta M_F = \pm 1$ transitions, for example by magnetic resonance in an atomic beam, at a magnetic field high enough for the term $(g_J\mu_B B)^2/A$ to become significant now yields approximate information about the value of A, whereas at lower fields where the energy states depended linearly on B the transition frequency was independent of A. Such measurements have been used to estimate previously unknown values of A with precision sufficient to render a subsequent search for $\Delta F = \pm 1$ transitions at zero field (or very low field) successful. The latter transitions lead to a precise determination of the zero-field hyperfine structure intervals. For example, from eq. (9.69) the transition $F \to F - 1$, $\Delta M_F = 0$, has the frequency

$$h\nu = AF + (g_F - g_{F-1})\,\mu_B B M_F, \tag{9.74}$$

which yields A with a small correction for the effect of the magnetic field. For the transition in which $M_F = 0$ there is no magnetic field correction in first order.

The selection rules for magnetic dipole radiation follow from the general considerations of section 7.4. There is no change of parity, so transitions can take place *within* a hyperfine structure multiplet; and in weak field $\Delta F = 0$, ± 1, $F = 0 \nrightarrow F = 0$, $\Delta M_F = 0$, ± 1. The frequency is so low (radiofrequency) that spontaneous emission can be completely neglected. Emission and absorption are induced by an oscillating magnetic field. The polarization is such that for an oscillating magnetic field parallel to the z-axis, defined by the direction of the steady field B, $\Delta M_F = 0$

9.6. Zeeman effect of hyperfine structure

(σ polarization), and for an oscillating magnetic field perpendicular to the z-axis $\Delta M_F = \pm 1$ (π polarization). The nomenclature σ and π is opposite to that for electric dipole radiation because the oscillating magnetic field in an electromagnetic wave is perpendicular to the electric field.

The criterion for a strong field is $g_J \mu_B B \gg A$. Then, in eq. (9.64), $g_J \mu_B J_z B$ is the largest term and M_J becomes a good quantum number. But since the complete Hamiltonian of eq. (9.63) commutes with $I_z + J_z = F_z$, $M = M_I + M_J = M_F$ is always a good quantum number ($I_z + J_z = F_z$ is a constant of the motion) whatever the magnitude of the field—see the analogous eq. (8.26). Hence M_I is also a good quantum number even though the term $-\boldsymbol{\mu}_I \cdot \mathbf{B}$ may not be large compared with $A\mathbf{I} \cdot \mathbf{J}$. The representation appropriate to strong field is then $|IJM_IM_J\rangle$, and in first order the Zeeman terms give an energy shift

$$\Delta E = \langle IJM_IM_J| \, g_J\mu_B J_z B - g'_I\mu_B I_z B \, |IJM_IM_J\rangle$$
$$= g_J\mu_B B M_J - g'_I\mu_B B M_I. \tag{9.75}$$

In first order the $A\mathbf{I} \cdot \mathbf{J}$ term contributes

$$\Delta E = \langle IJM_IM_J| \, A \{I_zJ_z + \tfrac{1}{2}(I_+J_- + I_-J_+)\} \, |IJM_IM_J\rangle$$
$$= AM_IM_J \tag{9.76}$$

to the energy shift, so the total energy shift is

$$\Delta E = g_J\mu_B B M_J - g'_I\mu_B B M_I + AM_IM_J \tag{9.77}$$

analogous to eq. (8.22).

In the limit of small g'_I the selection rules for magnetic dipole radiation are $\Delta M_I = 0, \Delta M_J = \pm 1$, which we can refer to as electron resonance in radiofrequency spectroscopy. The frequency for such transitions is given by

$$h\nu = g_J\mu_B B + AM_I. \tag{9.78}$$

Nuclear resonance—$\Delta M_J = 0, \Delta M_I = \pm 1$—is also possible, but in the limit of complete decoupling of $\mathbf{I}$ and $\mathbf{J}$ much greater radiofrequency power is required for nuclear resonance than for electron resonance. This is because the transition probability, which is proportional to

$$|\langle IJM_IM_J| \, g'_I\mu_B I_\pm(B_x \pm iB_y) \, |IJM_I \mp 1M_J\rangle|^2,$$

is comparable with that for electron resonance only if the rotating field $(B_x \pm iB_y)$ is g_J/g'_I ($\sim 10^3$) times as large as for electron resonance. This is equivalent to a factor of 10^6 in the radiofrequency power. The transition frequency in nuclear resonance is

$$h\nu = -g'_I\mu_B B + AM_J \tag{9.79}$$

which leads to a direct determination of g_I' if A is known. Such a determination is independent of the electronic radial parameter $\langle 1/r^3 \rangle$, and is the basis for much work both in free atoms and in the solid state.

A schematic energy level diagram for the example $J = \frac{3}{2}$, $I = 3$, which we discussed for weak field, can now be completed. For strong field, with $A > 0$, $g_I' \approx 0$, the energy levels are shown at the right of Fig. 9.7. The weak-field and strong-field levels can be connected unambiguously by following the rule that levels of the same $M = M_F = M_I + M_J$ never cross.

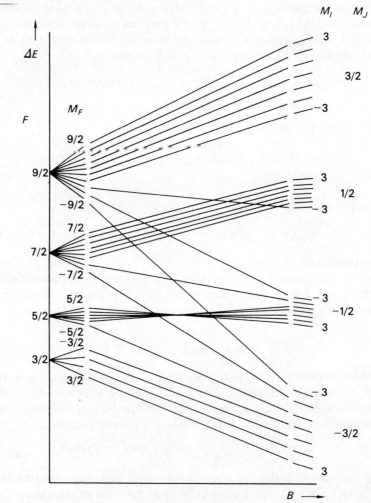

Fig. 9.7. Zeeman effect of hyperfine structure. $J = \frac{3}{2}$, $I = 3$, $A > 0$, $g_I' \approx 0$. At the left are the low-field states and at the right are the high-field states. In between, states of the same total M have simply been joined schematically by straight lines according to the rule that states with the same M never cross.

9.6. Zeeman effect of hyperfine structure

In the intermediate field region the energies of the states are given by the solution of a secular equation. Since $M = M_F = M_I + M_J$ represents a constant of the motion at all fields the matrix of the Hamiltonian (eq. (9.64)) breaks up into submatrices of given M. The case of $J = \frac{1}{2}$, arbitrary I, is an important simple one because it applies to the $^2S_{1/2}$ ground states of hydrogen and of the alkalis on which a great deal of work has been done. Since $M = M_I \pm \frac{1}{2}$ the largest secular determinant is 2×2 and the energy shift can be expressed in closed form. The result is the well-known Breit–Rabi formula:

$$\Delta E(F, M) = -\frac{h\,\Delta v}{2(2I + 1)} - g'_I \mu_B B M \pm \tfrac{1}{2} h\,\Delta v$$

$$\times \left\{ 1 + \frac{4M}{2I + 1} x + x^2 \right\}^{1/2} \quad (9.80)$$

where

$$x = \frac{(g_J + g'_I)\mu_B B}{h\,\Delta v} \quad (9.81)$$

and $h\,\Delta v$ is the separation between the levels $F = I + \frac{1}{2}$ and $F = I - \frac{1}{2}$ at zero field. The upper sign in eq. (9.80) is taken for states which belong

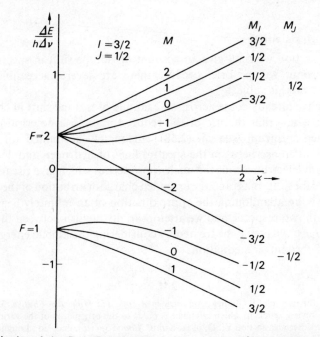

Fig. 9.8. A plot of the Breit–Rabi formula for the case $I = \frac{3}{2}$. The abscissa is $x = (g_J + g'_I)\,\mu_B B/(h\Delta v)$ where $h\Delta v$ is the energy difference between the levels $F = 2$ and $F = 1$ at zero field. This plot would apply, for example, to the $^2S_{1/2}$ ground level of ^{39}K. (See problem 9.5.)

at low field to $F = I + \frac{1}{2}$ and the lower sign for states which belong to $F = I - \frac{1}{2}$. The quadrupole interaction, which we omitted for convenience in our earlier discussion, vanishes for $J = \frac{1}{2}$, so there is no approximation in the Breit–Rabi formula in this respect. If the small term $g'_I \mu_B B M$ is omitted, the use of the parameter x as a measure of B allows one to make a universal plot of $\Delta E/(h \, \Delta v)$ against x. Such a plot is shown in Fig. 9.8 for $I = \frac{3}{2}$.

The Zeeman effect of hyperfine structure is important in magnetic resonance methods of spectroscopy, as mentioned earlier. Of these, the method of atomic beams has had particular success in the study of ground states and metastable states of atoms, including short-lived radioactive atoms. For excited states, on the other hand, very little work has been done with atomic beams, and much of the data has been obtained from optical spectroscopy. Recently, however, the methods of optical pumping, double resonance, and level-crossing spectroscopy† have provided data on the hyperfine structure of excited states, and more recently still, the techniques of Doppler-free laser spectroscopy have begun to be used. All this work is of interest not only because it has provided measurements of nuclear properties but also because it has stimulated a renewed effort to understand the way in which calculations on atomic structure should be made.

9.7. Isotope shift

In this section we shall give an account of isotope shift in spectral lines. The account will be brief because there are several accessible review articles‡ on this subject.

We have already come across an example of isotope shift in chapter 2, where we saw that the use of the reduced mass in a description of the hydrogen spectrum (see eq. (2.58)) adequately accounted for the frequency difference between the spectral lines of hydrogen and deuterium. For many-electron atoms the nuclear properties which give rise to isotope shift are the finite mass and the extended charge distribution of the nucleus. Thus we are abandoning the approximation of an infinitely heavy point charge in two respects, and we attempt to distinguish between the two.

The *mass effect* may be treated by considering the kinetic energy operator in Schrödinger's equation:

$$T = \frac{\mathbf{p}_n^2}{2M} + \sum_i \frac{\mathbf{p}_i^2}{2m_0} \tag{9.82}$$

† See, for example, B. Budick, *Advances in Atomic and Molecular Physics*, 3, 73, 1967. This is a review article in which reference is made to the originators of the various optical techniques. See also section VI. D. of H. Kuhn, *Atomic Spectra*, 2nd. Ed., Longmans, 1970.

‡ See, for example, H. Kopfermann, *Nuclear Moments*, Academic Press, 1958, chapter 1, section IV; G. Breit, *Rev. Mod. Phys.* **30**, 507, 1958; A. R. Bodmer, *Nucl. Phys.* **9**, 371, 1959; and, more recently, D. N. Stacey, *Reports on Progress in Physics*, **XXIX**, 171, 1966; H. Kuhn, *Atomic Spectra*, 2nd Ed., Longmans, 1970, section VI C.

9.7. Isotope shift

where $\mathbf{p}_n$ and M are the momentum and mass of the nucleus, and $\mathbf{p}_i$ and m_0 are the momentum and mass of the ith electron. From conservation of momentum for a stationary atom

$$\mathbf{p}_n = -\sum_i \mathbf{p}_i \qquad (9.83)$$

and so eq. (9.82) may be re-written

$$T = \frac{(\sum_i \mathbf{p}_i)^2}{2M} + \frac{\sum_i \mathbf{p}_i^2}{2m_0}$$

$$= \frac{\sum_i \mathbf{p}_i^2}{2M} + \frac{1}{M} \sum_{i>j} \mathbf{p}_i \cdot \mathbf{p}_j + \frac{\sum_i \mathbf{p}_i^2}{2m_0}. \qquad (9.84)$$

If at first we ignore the second term in eq. (9.84) which contains the dot product of electron momenta, the remaining two terms can be combined with the introduction of the reduced mass $m = m_0 M/(m_0 + M)$. Explicitly, an energy level $E(M)$ for an atom whose nucleus has a finite mass M is raised above the fictitious level $E(\infty)$ for an atom whose nucleus is infinitely heavy:

$$E(M) = E(\infty) \frac{M}{M + m_0} \qquad (9.85)$$

just as in eq. (2.58). We choose $E(\infty)$ as a reference level and use the notation ΔE for $E(M) - E(\infty)$, so

$$\Delta E = -E(\infty) \frac{m_0}{M + m_0}. \qquad (9.86)$$

For two isotopes whose nuclear masses, M and M', differ by δM we use the notation $\delta(\Delta E)$ for energy difference:

$$\delta(\Delta E) \approx E(\infty) \frac{m_0 \, \delta M}{MM'}. \qquad (9.87)$$

Finally, for the observed spectral line the wavenumber shift is

$$\delta \tilde{v} \approx \frac{m_0 \, \delta M}{MM'} \tilde{v} \qquad (9.88)$$

where the isotope of *greater* mass has the *larger* wavenumber. This phenomenon is known as the *normal mass shift*. The fractional shift $\delta \tilde{v}/\tilde{v}$ is easily calculated: it is largest for hydrogen–deuterium for which $\delta \tilde{v}/\tilde{v} = 2 \cdot 7 \times 10^{-4}$. Because of the proportionality to $1/M^2$ its effect is small in heavy elements.

Inclusion of the cross term in eq. (9.84) leads to an additional isotope shift in spectral lines known as the *specific mass shift*. This term involves a correlation between the motions of a pair of electrons, and the effect is very difficult to calculate for many-electron atoms. Like the normal shift,

195

the specific mass shift falls off as $1/M^2$ with increasing M, and is very small in the heaviest elements. Unfortunately the mass effects are not negligible in the intermediate region of $Z = 50$, and although some attempts have been made to estimate the specific mass shift in this region most of the work has been devoted to calculations for very light elements.

As a first approximation one can treat the operator

$$(1/M) \sum_{i>j} \mathbf{p}_i \cdot \mathbf{p}_j = (-\hbar^2/M) \sum_{i>j} \nabla_i \cdot \nabla_j$$

by perturbation theory. This leads to an energy shift in first order (again, the reference level is E(∞))

$$\Delta E = -\frac{\hbar^2}{M} \sum_{i>j} \int \psi^* \nabla_i \cdot \nabla_j \psi \, d\tau \qquad (9.89)$$

where ψ is the many-electron wave function. From this one next needs to find $\delta(\Delta E)$ for two isotopes and hence $\delta \tilde{\nu}$ for spectral lines. We shall not be able to discuss this mass effect further, except to remark that it is clear that the success of the calculation depends on the accuracy of the wave function!

Isotope shifts arising from the variation of the charge distribution of the nucleus from one isotope to another (sometimes called *field effects*) have received much attention, because the measurement of these shifts in spectral lines enables one to study the size and shape of nuclei as a function of neutron number. A good reason for wanting to estimate mass effects, especially the troublesome specific mass effect, is that the mass shifts have to be subtracted from the experimental results if the pure nuclear charge effects are to be investigated.

To treat the field effects we go back to the electric monopole interaction between the nucleus and the electrons of an atom, and consider the departure, for real nuclei, from the pure Coulomb potential of a nuclear point charge. We need a probe of the nuclear electric charge distribution: that is, we need to consider *s*-electrons,† whose charge density does not vanish in the region of the nucleus. Then we can expect an isotope shift in a spectral line for which the number of *s*-electrons is different in the two terms involved in the transition.

Early in the history of isotope shift it was realized that a relativistic treatment is necessary for evaluating the *s*-electron density $|\psi(0)|^2$ at the nucleus. Further, a first-order perturbation treatment is not really adequate: for in that theory one calculates ΔE, the energy shift of a term arising from the difference in electrostatic potential between a finite nucleus and a fictitious point nucleus, on the assumption that $|\psi(0)|^2$ is the zeroth-order electron density appropriate to a point nucleus. This

† In a relativistic treatment, $p_{\frac{1}{2}}$ electrons also have a non-vanishing charge density near the origin. Other electrons with $l \neq 0$, however, can be regarded as probing the nucleus to a negligible extent.

approximation is not very accurate because the finite nuclear charge distribution causes a significant modification to $|\psi(0)|^2$.

Notwithstanding these important qualifications, we shall give a brief account of the non-relativistic, first-order perturbation treatment to obtain an order of magnitude of the field effects in heavy elements. We assume for simplicity a nuclear model in which the charge distribution is spherically symmetrical with radius r_0 given by

$$r_0 = 1 \cdot 2 \times 10^{-13} A^{1/3} \text{ cm} \tag{9.90}$$

where A is the mass number; and we consider the isotope shift of a term for two isotopes whose nuclear radii differ by

$$\frac{\delta r_0}{r_0} = \frac{1}{3} \cdot \frac{\delta A}{A}. \tag{9.91}$$

This particular field effect is called the *volume shift*. In first order the contribution to the energy shift of a term from each electron is the expectation value of the electrostatic potential energy difference $V(r) - V_0(r)$, where $V(r)$ is the potential energy appropriate to an extended nucleus and $V_0(r)$ is that appropriate to a point nucleus. Thus

$$\Delta E = \int_0^\infty \psi^*(V(r) - V_0(r))\psi 4\pi r^2 \, dr$$

$$\approx |\psi(0)|^2 \int_0^{r_0} (V(r) - V_0(r)) 4\pi r^2 \, dr. \tag{9.92}$$

In this equation we have restricted the range of integration to $0 \leqslant r \leqslant r_0$ since, by Gauss's theorem, $V(r) = V_0(r)$ for $r \geqslant r_0$; and we have assumed that over this range the electron charge density is approximately constant and has the value $-e|\psi(0)|^2$. For a uniform nuclear charge distribution it can easily be shown that (see problem 2.9(e))

$$V(r) = \frac{Ze^2}{4\pi\varepsilon_0 r_0}\left(-\frac{3}{2} + \frac{1}{2} \cdot \frac{r^2}{r_0^2}\right), \quad 0 \leqslant r \leqslant r_0 \tag{9.93}$$

whereas $V_0(r) = -Ze^2/4\pi\varepsilon_0 r$, so ΔE becomes

$$\Delta E = \frac{4\pi}{10} |\psi(0)|^2 \frac{Ze^2}{4\pi\varepsilon_0} r_0^2. \tag{9.94}$$

On the assumption that all the electrons move independently in the central field of the nucleus, this expression for ΔE should be summed over all s-electrons. But since we are dealing with a spectral *line* we need only sum over those valence s-electrons by which the two terms involved in the transition differ.

The quantity which is directly related to observation is the difference in ΔE for two isotopes:

$$\delta(\Delta E) = \frac{4\pi}{5} |\psi(0)|^2 \frac{Ze^2}{4\pi\varepsilon_0} r_0^2 \frac{\delta r_0}{r_0}. \tag{9.95}$$

It is conventional to re-write this in the following form

$$\delta(\Delta E) = |\psi(0)|^2 \frac{\pi a_0^3}{Z} \times \frac{2}{5} R_\infty \left(\frac{2Zr_0}{a_0}\right)^2 \frac{\delta r_0}{r_0} \tag{9.96}$$

which can be written in the more general case

$$\delta(\Delta E) = |\psi(0)|^2 \frac{\pi a_0^3}{Z} C\left(Z, r_0, \frac{\delta r_0}{r_0}\right). \tag{9.97}$$

This expression is a product of two factors: the first factor depends, through $|\psi(0)|^2$, on the electron charge distribution and the second factor, C, is a function of the difference between nuclear charge distributions for two isotopes.

In view of the simplifications we have made, the non-relativistic first-order expression (9.96) for the volume shift gives no better than an order of magnitude estimate. Nevertheless the form of eq. (9.97) is satisfactory, and refinements can be introduced by modification of the factor C. Equation (9.96) indicates that the isotope with larger r_0 has the higher energy value and this is in agreement with experiment. Further, the expression suggests that the variation in the nuclear mean square radius, $\langle r^2 \rangle$, between isotopes is an important quantity for investigation. Not only differences in nuclear *size*, but also differences in nuclear *shape* contribute to $\delta\langle r^2 \rangle$. Thus it is of interest for nuclear structure to investigate a run of C-factors along a series of neighbouring isotopes, and some work has been done with this in view. As the neutron number is increased by two in a series of even-even isotopes the lines for a given element are ordered according to mass number. But the lines are not necessarily equally spaced in frequency as would be expected from the model assumed in eq. (9.91), and these departures from equal spacing are significant in the study of nuclear charge distribution. In particular a line of an isotope with odd mass number (or rather the centre of gravity of the hyperfine structure components of such a line) does not lie half-way between the lines of its even-even neighbours, but is displaced towards the line of the lighter isotope. The tentative explanation of this phenomenon, called odd-even staggering, is beyond the scope of this discussion (but see, for example, the above-mentioned review article by Stacey who refers to the work of H. Stroke).

To give an idea of the order of magnitude of experimental shifts we quote two examples of isotope shift measurements on elements in the middle of the periodic table—for heavy elements the shifts are larger.

Problems

Hindmarsh and Kuhn† have examined the transition $5s^2 6s \, ^2S_{1/2}$—$5s^2 6p \, ^2P_{3/2}$ in Sn II (tin has many stable isotopes). They chose this transition with the expectation that the $6s$ electron would be responsible for a reasonable differential shift of the $^2S_{1/2}$ level. The order of magnitude of the shift between even-even isotopes is 5 mK which is small enough compared with the line-width to necessitate the use of enriched isotopes, a procedure which has obvious advantages and is common practice in isotope shift work. The second example which we quote is the work of Kuhn and Ramsden‡ on the transition $4d^{10} 5p \, ^2P_{3/2}$—$4d^9 5s^2 \, ^2D_{5/2}$ in Cd II. The isotope shift for the addition of two neutrons is in this case about 50 mK. This transition involves a two-electron jump, and the contribution of the two equivalent $5s^2$ electrons to the isotope shift is expected to be twice as large as might have been the case if only one s-electron had jumped. This example brings out the distinction between the isotope shift, which arises from an *electrostatic* interaction, and the magnetic hyperfine structure, for which the $5s^2$ electrons, paired off with respect to spin orientation, would give zero *magnetic* field at the nucleus. The hyperfine structure of this transition in the odd isotopes of Cd is in fact due largely to the d^9 configuration, and is of the same order of magnitude as the isotope shifts.

Finally, we mention that there are other spectroscopic ways of providing a probe of the nucleus with which to study nuclear charge distribution and isotope shift. One of these, the spectroscopy of muonic atoms, can be contrasted with optical spectroscopy as follows: whereas we have been speaking of the differential shift $\delta(\Delta E)$ of an energy level, which is the quantity deduced from the line shifts observed in optical spectra, one can derive ΔE itself from a study of muonic X-rays. The reason for this is that the energy levels $E(\infty)$ for a fictitious muonic atom containing a point nucleus can be calculated: the problem is a two-body one with certain corrections. In particular the specific mass effect is absent. Thus ΔE rather than $\delta(\Delta E)$ is obtained directly from experiment. There is a need for systematic data in these studies of nuclear charge distribution, and it is hoped that the developing field of measurement of isotope shifts by muonic spectroscopy will complement the work in optical spectroscopy.

Problems

9.1. Derive eq. (9.24) from purely classical considerations.

9.2. For hydrogen in a non-relativistic approximation evaluate the following ratios of the hyperfine structure magnetic dipole splitting factor:

$$a(n \, ^2S_{1/2}) : a(n \, ^2P_{1/2}) : a(n \, ^2P_{3/2}) : a(n \, ^2D_{3/2}) : a(n \, ^2D_{5/2}).$$

† W. R. Hindmarsh and H. Kuhn, *Proc. Phys. Soc. A* **68**, 433, 1955.
‡ H. G. Kuhn and S. A. Ramsden, *Proc. Roy. Soc. A* **237**, 485, 1956.

9.3. Consider the hyperfine structure of the transition $5d^26s\ ^4F_{9/2}$—$5d^26p\ ^4G_{11/2}$ in ^{139}La.

(a) Why might you suspect that the magnetic dipole hyperfine structure of the $^4F_{9/2}$ level is considerably larger than that of the $^4G_{11/2}$ level?

(b) Would you expect an electric quadrupole interaction to give a striking departure from the interval rule in the $^4F_{9/2}$ level?

(c) In what follows assume that the hyperfine structure of the $^4G_{11/2}$ level is negligibly small and that the electric quadrupole interaction can also be neglected. Those hyperfine structure components which are well resolved in this transition lie at 0, 111·0, 208·0, 291·6, 361·3, 417·0, 459·2 mK above the lowest frequency component. By counting the number of components set a lower limit to the nuclear spin of ^{139}La.

(d) The limits of error on the frequency measurements of the components are about ± 0.2 mK. Determine the value of the nuclear spin from the frequency measurements.

(e) If a measurement of the relative intensities of the components were reliable, such a measurement would confirm the evaluation of the nuclear spin. What are the relative intensities expected to be?

(f) Given that the $^4F_{9/2}$ level lies below the $^4G_{11/2}$ level, determine the sign of the magnetic dipole splitting constant $A(^4F_{9/2})$. Also estimate the magnitude of $A(^4F_{9/2})$ from the frequency measurements.

(g) The transition has a wavelength of 6,250 Å. What would be the Doppler width of the hyperfine structure components if the light source were run at room temperature? Comment on this, and on the use of a Fabry–Perot etalon to resolve the hyperfine structure.

9.4. For the $3d^54s4p\ ^6P_{7/2}$ level of ^{55}Mn discussed in the text find expressions for the departures from the interval rule in terms of B, the electric quadrupole interaction constant. Then show that the value $B(^6P_{7/2}) = 2·2 \pm 0·4$ mK is consistent with the experimental data of table 9.1, column 4.

9.5. A magnetic resonance experiment by the method of atomic beams has been done to find the nuclear spin of ^{40}K, whose ground level is known to be $^2S_{1/2}$ with $g_J = 2·0023$. In this experiment a magnetic dipole transition $\Delta F = 0$, $\Delta M_F = \pm 1$ is observed in a weak magnetic field: the particular transition is $F = F_{max} = I + J$, with $M_F = -(I + J - 1) \leftrightarrow M_F = -(I + J)$, where I is to be found. (This transition is in fact the only one with $\Delta F = 0$ which can be observed in a conventional apparatus when $J = \frac{1}{2}$.) The frequency of the transition is 1·557 MHz in a certain field B which is evaluated by measuring the analogous transition frequency with an atomic beam of ^{39}K for which I is known to be $\frac{3}{2}$: $v(^{39}K) = 3·504$ MHz.

(a) What is the magnetic field B in gauss?

(b) What is the nuclear spin of ^{40}K?

(c) Estimate the precision to which the frequencies and g_J need to be known in order to distinguish the result for $I(^{40}\text{K})$ from neighbouring possibilities.

(d) In a subsequent experiment the separation between the two F levels in zero field is found to be 1,285·79 MHz for ^{40}K. What is the magnetic dipole splitting constant $A(^2\text{S}_{1/2})$ for ^{40}K? Can its sign be determined from this frequency measurement?

(e) In a high field representation there is a group of states with $M_J = \frac{1}{2}$ and various values of M_I. How many such states are there? Estimate the separation in energy between these M_I states in a high field of (say) 0·5 T. (Neglect the direct interaction of the nuclear magnetic moment with the external field.)

(f) The nuclear moment of ^{40}K is about $-1\cdot3$ nuclear magnetons. By how much does the direct interaction of the nuclear moment with the external field shift each of the energy states of part (e) in a field of 0·5 T?

Appendix A

We summarize here without rigour some of the theorems of quantum mechanics which we shall want to use.

A.1. Eigenfunctions

An operator **A** representing an observable has a complete, normalized orthogonal set of eigenfunctions which are linearly independent. The word 'complete' means that there is no other eigenfunction of **A** which is not a linear combination of members of the set. We confine ourselves to operators **A** which have a discrete set of eigenvalues a_i.

A.2. Degeneracy

If there are m eigenfunctions $u_1, u_2, \ldots, u_m$ of **A** corresponding to the same eigenvalue, the system is said to be m-fold degenerate.

A.3. Expansion theorem

Any physically acceptable wave function can be expanded in terms of a complete normalized orthogonal set of eigenfunctions u_i:

$$\Psi = \sum_i c_i u_i \quad \text{where} \quad \int u_j^* u_i \, d\tau = \delta_{ji}. \tag{A.1}$$

If u_i is an eigenfunction of **A** with eigenvalue a_i, i.e.,

$$\mathbf{A} u_i = a_i u_i \tag{A.2}$$

then the expectation value of **A** in the state Ψ is

$$\begin{aligned}
\langle A \rangle_\Psi &= \int \Psi^* \mathbf{A} \Psi \, d\tau, \\
&= \sum_j \sum_i \int c_j^* u_j^* \mathbf{A} c_i u_i \, d\tau, \\
&= \sum_j \sum_i c_j^* c_i a_i \int u_j^* u_i \, d\tau, \\
&= \sum_j \sum_i c_j^* c_i a_i \delta_{ji}, \\
&= \sum_i c_i^* c_i a_i.
\end{aligned} \tag{A.3}$$

This means that if we make a measurement of **A** on a system in the state Ψ the result will be a_i with probability $|c_i|^2$, where $\sum_i |c_i|^2 = 1$ if Ψ is normalized. Thus, the expansion is very like a Fourier expansion. In particular, if Ψ is identical with u_i the coefficients $c_{j \neq i}$ are zero. Then the result of a measurement of **A** is, with certainty, the eigenvalue a_i.

A.4. Matrix elements

Consider an operator **B** for which u_i is not an eigenfunction. The expectation value of **B** in the state Ψ, where $\Psi = \sum_i c_i u_i$, is

$$\langle \mathbf{B} \rangle_\Psi = \int \Psi^* \mathbf{B} \Psi \, d\tau$$

$$= \sum_j \sum_i c_j^* \left(\int u_j^* \mathbf{B} u_i \, d\tau \right) c_i \tag{A.4}$$

which we shall write in the so-called Dirac notation

$$\langle \mathbf{B} \rangle_\Psi = \sum_j \sum_i c_j^* \langle j| \, \mathbf{B} \, |i\rangle c_i. \tag{A.5}$$

Thus, $\langle j| \, \mathbf{B} \, |i\rangle$ is the j–ith element of a square matrix. Matrix elements depend not only on the operator (**B**) but also on the wave-functions used to compute them (u_i, written $|i\rangle$). The particular choice of the set u_i used in the expansion of Ψ is called the *basis set* of the *representation* u_i. Whereas we introduced an expectation value as a kind of quantum-mechanical average, we now see that off-diagonal matrix elements arise as a direct consequence of the principle of superposition embodied in the expansion theorem. They are the *interference terms*.

A.5. Hermitian operators

We can reasonably demand that operators which represent actual dynamical quantities should give real expectation values. This is guaranteed if such operators are *Hermitian*, that is, if they obey

$$\int u_j^* \mathbf{B} u_i \, d\tau = \int (\mathbf{B} u_j)^* u_i \, d\tau \tag{A.6}$$

for then

$$\langle \mathbf{B} \rangle = \int u_i^* \mathbf{B} u_i \, d\tau$$

and

$$\langle \mathbf{B} \rangle^* = \int u_i (\mathbf{B} u_i)^* \, d\tau$$

$$= \int (\mathbf{B} u_i)^* u_i \, d\tau$$

$$= \int u_i^* \mathbf{B} u_i \, d\tau \quad \text{from (A.6)}$$

$$= \langle \mathbf{B} \rangle. \qquad (A.7)$$

Hence, $\langle \mathbf{B} \rangle$ is real.

In the Dirac notation (A.6) is

$$\langle j | \, \mathbf{B} \, | i \rangle = \langle i | \, \mathbf{B} \, | j \rangle^*,$$

that is, a matrix element of a Hermitian operator is the complex conjugate of the transposed matrix element. In particular the diagonal elements are real:

$$\langle i | \, \mathbf{B} \, | i \rangle = \langle i | \, \mathbf{B} \, | i \rangle^*. \qquad (A.8)$$

An eigenvalue problem is the case when the representation is chosen such that off-diagonal elements vanish:

$$\langle j | \, \mathbf{A} \, | i \rangle = a_i \delta_{ji}. \qquad (A.9)$$

A.6. Commuting operators

If $\mathbf{A}$ represents a real observable with non-degenerate eigenfunctions u_i (or $|i\rangle$), that is, if

$$\mathbf{A} | i \rangle = a_i | i \rangle \text{ with } \langle j | \, \mathbf{A} \, | i \rangle = a_i \delta_{ji}, \qquad (A.10)$$

then for another operator $\mathbf{B}$

$$\langle k | \, \mathbf{BA} \, | i \rangle = \langle k | \, \mathbf{B} \, | i \rangle a_i. \qquad (A.11)$$

Also if we expand a new function $\mathbf{B} | i \rangle$ in terms of the set u_j (or $|j\rangle$):

$$\mathbf{B} | i \rangle = \sum_j c_j | j \rangle, \qquad (A.12)$$

where

$$\langle j | \, \mathbf{B} \, | i \rangle = c_j \qquad (A.13)$$

from the orthogonality of the $|j\rangle$, then

$$\langle k| \mathbf{AB} |i\rangle = \sum_j \langle k| \mathbf{A} |j\rangle \langle j| \mathbf{B} |i\rangle, \qquad (A.14)$$

$$= a_k \langle k| \mathbf{B} |i\rangle.$$

Hence,

$$\langle k| \mathbf{BA} - \mathbf{AB} |i\rangle = \langle k| \mathbf{B} |i\rangle (a_i - a_k). \qquad (A.15)$$

But by hypothesis all the a_i are different, so if $\mathbf{B}$ and $\mathbf{A}$ commute, i.e., if $[\mathbf{B}, \mathbf{A}] \equiv \mathbf{BA} - \mathbf{AB} = 0$, from (A.15)

$$\langle k| \mathbf{B} |i\rangle = 0 \text{ unless } k = i. \qquad (A.16)$$

This means that in a representation in which $\mathbf{A}$ is diagonal $\mathbf{B}$ is also diagonal if $\mathbf{B}$ commutes with $\mathbf{A}$. $\mathbf{B}$ and $\mathbf{A}$ have simultaneous eigenfunctions.

Appendix B

We quote here the results of time-independent perturbation theory in first and second order.

We assume that we start with a zeroth-order eigenvalue equation

$$\mathcal{H}_0 u_{0m} = E_{0m} u_{0m}, \tag{B.1}$$

where u_{0m} is the eigenfunction belonging to the eigenvalue E_{0m} for the state m of the zeroth-order Hamiltonian $\mathcal{H}_0$. We shall write the zeroth-order eigenfunctions u_{0m} simply as $|m\rangle$, so eq. (B.1) becomes

$$\mathcal{H}_0 |m\rangle = E_{0m} |m\rangle. \tag{B.2}$$

We now consider the effect of adding a small perturbation V to the zeroth-order Hamiltonian $\mathcal{H}_0$ to give

$$\mathcal{H} = \mathcal{H}_0 + V. \tag{B.3}$$

We are chiefly interested to know how the energies are modified as a result of the perturbation. We have to distinguish two cases.

B.1. Non-degenerate case

It is assumed that all the E_{0m} are different, that is, the states $|m\rangle$ are non-degenerate. Then the first-order correction to the energy E_{0m} for the state m is

$$\Delta E_{1m} = \langle m| V |m\rangle, \tag{B.4}$$

which is just the expectation value of the perturbation V taken with the unperturbed functions.

The second-order correction is

$$\Delta E_{2m} = \sum_k{}' \frac{\langle m| V |k\rangle \langle k| V |m\rangle}{E_{0m} - E_{0k}}, \tag{B.5}$$

where the summation is taken over all the states of the complete set which satisfies eq. (B.2), except the state m itself. The omission of the state m in the sum is indicated by the prime on the summation sign. The energy of the state m, up to second order, is

$$E_m = E_{0m} + \Delta E_{1m} + \Delta E_{2m}. \tag{B.6}$$

The label m is that appropriate to zeroth-order, and all the matrix elements of V are taken with zeroth-order functions. The wave function itself is modified by the perturbation. To first order it becomes

$$\Psi_{1m} = |m\rangle + \sum_k{}' |k\rangle \frac{\langle k| \, V \, |m\rangle}{E_{0m} - E_{0k}}. \tag{B.7}$$

That is, the wave function is no longer the pure state m, but it now has some of the character of the other states k of the set. One says that the state k is mixed into the state m with amplitude

$$a_{km} = \frac{\langle k| \, V \, |m\rangle}{E_{0m} - E_{0k}}.$$

Clearly the quantum numbers designated by the label m are no longer strictly good quantum numbers, but one retains the label m for Ψ on the grounds that the amplitude of admixture a_{km} has been assumed to be small.

B.2. Degenerate case

If some of the states u_{0k} are degenerate with u_{0m}, $E_{0m} - E_{0k}$ vanishes in the denominator of eq. (B.5), and one has trouble if $\langle m| \, V \, |k\rangle \neq 0$. This difficulty can be overcome by choosing a different set of zeroth-order functions. This procedure emphasizes the arbitrary nature of the choice of *representation* u_{0m}. One naturally chooses the representation best adapted to the particular problem. If the first n functions $u_{0l}(l = 1, \ldots, n)$ are n-fold degenerate, all with energy E_{0m}, we can start again and find a new set of normalized orthogonal functions v_{0j} which are linear combinations of the u_{0l}:

$$v_{0j} = \sum_{l=1}^{n} c_{jl} u_{0l}, \tag{B.8}$$

with the amplitudes c_{jl} chosen in such a way that

$$\langle v_{0i}| \, V \, |v_{0j}\rangle = 0, \quad i \neq j, \tag{B.9}$$

in the new v-representation. The new set of v_{0j} are still n-fold degenerate eigenfunctions of $\mathcal{H}_0$ because

$$\mathcal{H}_0 v_{0j} = \sum_{l=1}^{n} c_{jl} \mathcal{H}_0 u_{0l},$$

$$= \sum_{l=1}^{n} c_{jl} E_{0m} u_{0l} \quad \text{from (B.1)}$$

$$= E_{0m} v_{0j} \qquad \text{from (B.8)} \tag{B.10}$$

but now the troublesome terms in eq. (B.5) drop out. The choice of the expansion coefficients c_{jl} which achieves this result depends on the form of V through eq. (B.9): a diagonalization of the n-fold sub-matrix of V in the u_{0l} representation serves to evaluate the coefficients c_{jl} and hence to define the functions v_{0j} through eq. (B.8). The first-order energy shift for one of the n degenerate states is then a diagonal matrix element of V in the v-representation:

$$\Delta E_{1j} = \langle v_{0j} | \, V \, | v_{0j} \rangle \tag{B.11}$$

rather than eq. (B.4).

The case of degeneracy occurs frequently in the theory of atomic structure. See, for example, how degenerate perturbation theory is applied to the fine structure of hydrogen (p. 68), the electrostatic interaction in helium (p. 77), the spin-orbit interaction in LS coupling (p. 125), the Zeeman effect in LS coupling (p. 147), and the magnetic-dipole hyperfine structure splitting (p. 171).

Appendix C

Notes on angular momentum; a summary of results.†

C.1. Orbital angular momentum

The classical definition of the angular momentum of a single particle about the origin of co-ordinates is

$$\mathscr{L} = \mathbf{r} \times \mathbf{p}. \tag{C.1}$$

In quantum mechanics this becomes an operator which is obtained by the replacement $\mathbf{p} \rightarrow -i\hbar\nabla$:

$$\hbar\mathbf{l} = -i\hbar\mathbf{r} \times \nabla \tag{C.2}$$

where we have introduced the factor $\hbar$ into the definition of $\mathbf{l}$ in order to avoid carrying units of $\hbar$.

The most important point about $\mathbf{l}$ is that its components do not commute: e.g.,

$$[l_x, l_y] \equiv l_x l_y - l_y l_x \neq 0,$$

but

$$[l_x, l_y] = il_z, \quad \text{and cyclically.} \tag{C.3}$$

When there is more than one particle we can define a total orbital angular momentum operator

$$\mathbf{L} = \sum_i \mathbf{l}(i) \tag{C.4}$$

which also satisfies the commutation relation (C.3).

C.2. General definition of angular momentum

In order to include intrinsic spin angular momentum in our discussion we must generalize the definition of angular momentum. We consider as an

---

† See, for example, R. H. Dicke and J. P. Wittke, *Introduction to Quantum Mechanics*, chapter 9.

angular momentum any vector observable $\hbar\mathbf{J}$ if it satisfies the extension of (C.3):

$$J_x J_y - J_y J_x = iJ_z$$
$$J_y J_z - J_z J_y = iJ_x \qquad \text{(C.5)}$$
$$J_z J_x - J_x J_z = iJ_y$$

or in a condensed formal way

$$\mathbf{J} \times \mathbf{J} = i\mathbf{J}. \qquad \text{(C.6)}$$

In order that $\mathbf{J}$ be an observable it must be Hermitian (see eq. (A.6)). In general one introduces the Hermitian conjugate $\mathbf{B}^\dagger$ of an operator $\mathbf{B}$ where $\mathbf{B}^\dagger$ is defined by the process of taking the complex conjugate of a matrix element:

$$\int u_i^* \mathbf{B}^\dagger u_j \, d\tau = \left[\int u_j^* \mathbf{B} u_i \, d\tau \right]^*, \qquad \text{(C.7)}$$

that is, the matrix element of $\mathbf{B}\dagger$ is the complex conjugate of the transpose of the matrix element of $\mathbf{B}$. For a Hermitian operator $\mathbf{B}^\dagger = \mathbf{B}$. We mention this because we find it convenient to use two *non-Hermitian* operators

$$J_+ = J_x + iJ_y \qquad \text{(C.8)}$$

and

$$J_- = J_x - iJ_y$$

to replace J_x and J_y. These have the property

$$J_+^\dagger = J_-$$
$$J_-^\dagger = J_+ \qquad \text{(C.9)}$$

and the commutation relations (C.5) become

$$J_+ J_- - J_- J_+ = 2J_z$$
$$J_+ J_z - J_z J_+ = -J_+ \qquad \text{(C.10)}$$
$$J_- J_z - J_z J_- = J_-.$$

We also need to introduce the operator $\mathbf{J}^2 = \mathbf{J} \cdot \mathbf{J}$ which represents the square of the magnitude of the angular momentum $\mathbf{J}$. $\mathbf{J}^2$ commutes with all the components of $\mathbf{J}$:

$$[\mathbf{J}^2, \mathbf{J}] = 0. \qquad \text{(C.11)}$$

Making use of (C.8) we find that

$$\mathbf{J}^2 = J_z^2 + \tfrac{1}{2}(J_+ J_- + J_- J_+), \qquad \text{(C.12)}$$

Appendix C

from which, with eq. (C.10), we can derive the useful results

$$J_-J_+ = \mathbf{J}^2 - J_z(J_z + 1)$$
$$J_+J_- = \mathbf{J}^2 - J_z(J_z - 1).$$ (C.13)

C.3. The eigenvalues of $\mathbf{J}^2$ and J_z

Because the components of $\mathbf{J}$ do not commute we cannot find a function which is a simultaneous eigenfunction of all the components of $\mathbf{J}$ (that is, we cannot simultaneously diagonalize all the components of $\mathbf{J}$): see section A.6 of appendix A. However, $\mathbf{J}^2$ does commute with $\mathbf{J}$, so we can simultaneously diagonalize $\mathbf{J}^2$ and one component of $\mathbf{J}$: by convention we always consider the component J_z.

Let the eigenfunctions of $\mathbf{J}^2$ and J_z be $|a, M\rangle$ with eigenvalues a and M, i.e.,

$$\mathbf{J}^2|a, M\rangle = a|a, M\rangle,$$
$$J_z|a, M\rangle = M|a, M\rangle.$$ (C.14)

We choose the states so that they are orthonormal, i.e.,

$$\langle a', M'|a, M\rangle = \delta_{aa'}\delta_{MM'}.$$ (C.15)

To set limits on the values of a and M we consider the new state $J_+|a, M\rangle$. Its norm is

$$\langle a, M|J_+^\dagger J_+|a, M\rangle = \langle a, M|J_-J_+|a, M\rangle.$$

But it is a fundamental theorem that the norm of a state must be greater than or equal to zero:

$$\langle a, M|J_-J_+|a, M\rangle \geq 0.$$ (C.16)

Likewise

$$\langle a, M|J_+J_-|a, M\rangle \geq 0.$$ (C.17)

It follows that

$$a - M(M + 1) \geq 0,$$
$$a - M(M - 1) \geq 0,$$ (C.18)

and, hence

$$a \geq M^2.$$ (C.19)

Returning to the state $J_+|a, M\rangle$ we find

$$J_zJ_+|a, M\rangle = (M + 1)J_+|a, M\rangle,$$

and (C.20)

$$\mathbf{J}^2J_+|a, M\rangle = aJ_+|a, M\rangle.$$

211

Equations (C.20) show that the state $J_+|a, M\rangle$ is an eigenfunction of $\mathbf{J}^2$ and of J_z, differing from $|a, M\rangle$ in that the eigenvalue of J_z is greater by one while the eigenvalue of $\mathbf{J}^2$ is the same. Repeated operation with J_+ yields a series of states all with the same eigenvalue a but with values of M increasing by one each time. Similarly repeated operation with J_- gives values of M decreasing by one each time. For this reason J_+ and J_- are called raising and lowering operators.

But the generation of new eigenfunctions cannot go on for ever without violating (C.19). Therefore, there must be states which satisfy

$$J_+|a, M_{max}\rangle = 0,$$
$$J_-|a, M_{min}\rangle = 0. \tag{C.21}$$

It then follows that

$$a - M_{max}(M_{max} + 1) = 0,$$
$$a - M_{min}(M_{min} - 1) = 0. \tag{C.22}$$

Hence

$$M_{max} = -M_{min}. \tag{C.23}$$

But the values of M go in steps of one, so $M_{max} - M_{min}$ is an integer. Hence, $2M_{max}$ and $2M_{min}$ are each integers, so *all the M are either integer or half-integer*. We now call the maximum value of M by the name J, and from (C.22) we can write

$$a = J(J + 1). \tag{C.24}$$

We use J from now on, instead of a, as one of the labels for the eigen-functions and we write them $|J, M\rangle$. J is an integer or half-integer greater than or equal to zero. M takes values between J and $-J$ differing by unity and is an integer or half-integer depending on which J itself is. We rewrite eq. (C.14) in the form

$$\mathbf{J}^2|J, M\rangle = J(J + 1)|J, M\rangle,$$
$$J_z|J, M\rangle = M |J, M\rangle. \tag{C.25}$$

C.4. Matrix elements of the angular momentum operators

First of all,

$$\langle J'M'| J_z |JM\rangle = M\delta_{JJ'}\delta_{MM'}. \tag{C.26}$$

Now we consider the matrix elements of J_+ and J_-. In view of eq. (C.20) we can write

$$J_+|J, M\rangle = c_{M+1}|J, M + 1\rangle \tag{C.27}$$

where c_{M+1} is a constant to be determined. Taking the norm of both sides of eq. (C.27), we have

$$\langle J, M | J_- J_+ | J, M \rangle = c_{M+1}^* c_{M+1} \langle J, M+1 | J, M+1 \rangle$$
$$= c_{M+1}^* c_{M+1} \qquad (C.28)$$

since $|J, M+1\rangle$ is normalized. But

$$\langle J, M | J_- J_+ | J, M \rangle = J(J+1) - M(M+1), \qquad (C.29)$$

so

$$c_{M+1} = e^{i\delta} [J(J+1) - M(M+1)]^{1/2}. \qquad (C.30)$$

Choosing the phase $e^{i\delta} = +1$, and sticking to this convention throughout, we have

$$J_+ |J, M\rangle = [J(J+1) - M(M+1)]^{1/2} |J, M+1\rangle, \qquad (C.31)$$

or

$$\langle J, M+1 | J_+ | J, M \rangle = [J(J+1) - M(M+1)]^{1/2}. \qquad (C.32)$$

Similarly,

$$\langle J, M-1 | J_- | J, M \rangle = [J(J+1) - M(M-1)]^{1/2}. \qquad (C.33)$$

These formulae are easy to remember because, in (C.32), J_+ raises M to $M+1$, and on the right-hand side the product $M(M+1)$ appears. All other matrix elements of J_+ and J_- vanish, and in particular the matrix elements of J_+ are diagonal in J.

C.5. Two commuting angular momenta

If we introduce another angular momentum **I** all of whose components commute with all the components of **J**, then it is possible to have a function $|I, J, M_I, M_J\rangle$ which is a simultaneous eigenfunction of $\mathbf{I}^2$, $\mathbf{J}^2$, I_z, and J_z:

$$\begin{aligned}
\mathbf{I}^2 |I, J, M_I, M_J\rangle &= I(I+1)|I, J, M_I, M_J\rangle, \\
\mathbf{J}^2 |I, J, M_I, M_J\rangle &= J(J+1)|I, J, M_I, M_J\rangle, \\
I_z |I, J, M_I, M_J\rangle &= M_I |I, J, M_I, M_J\rangle, \\
J_z |I, J, M_I, M_J\rangle &= M_J |I, J, M_I, M_J\rangle.
\end{aligned} \qquad (C.34)$$

An operator which one comes across frequently is $\mathbf{I} \cdot \mathbf{J}$:

$$\begin{aligned}
\mathbf{I} \cdot \mathbf{J} &= I_z J_z + I_x J_x + I_y J_y \\
&= I_z J_z + \tfrac{1}{2}(I_+ J_- + I_- J_+).
\end{aligned} \qquad (C.35)$$

Its matrix elements are diagonal in I and J. In the representation of eq. (C.34) $I_z J_z$ has only matrix elements diagonal in M_I and M_J:

$$\langle I, M_I', J, M_J' | I_z J_z | I, M_I, J, M_J \rangle = M_I M_J \delta_{M_I' M_I} \delta_{M_J' M_J}. \qquad (C.36)$$

The only non-vanishing matrix element of I_+J_- is

$$\langle I, M_I + 1, J, M_J - 1 | I_+J_- | I, M_I, J, M_J \rangle =$$
$$[I(I + 1) - M_I(M_I + 1)]^{1/2}[J(J + 1) - M_J(M_J - 1)]^{1/2}. \quad \text{(C.37)}$$

Similarly

$$\langle I, M_I - 1, J, M_J + 1 | I_-J_+ | I, M_I, J, M_J \rangle =$$
$$[I(I + 1) - M_I(M_I - 1)]^{1/2}[J(J + 1) - M_J(M_J + 1)]^{1/2}. \quad \text{(C.38)}$$

Notice that I_+J_- does not commute with I_z or J_z separately so it does not have matrix elements diagonal in M_I and M_J. But I_+J_- *does* commute with $I_z + J_z$ which has the eigenvalue $M = M_I + M_J$, so M is conserved in eq. (C.37), that is, in the matrix element

$$(M_I + 1) + (M_J - 1) = M_I + M_J = M.$$

Problems on Appendix C

Some of the statements, given without proof in appendix C, depend on what has gone before. It is therefore advisable to work through the equations as follows:

1. Derive eq. (C.3). Use (C.2).
2. Show that (C.4) satisfies (C.3). First show that $[I(i), I(j)] = 0$ if $i \neq j$.
3. Derive eq. (C.10) from (C.5).
4. Verify (C.11).
5. Derive (C.13).
6. Derive eq. (C.18).
7. Derive (C.19).
8. Derive (C.20).
9. Derive (C.22) and (C.23). Why must we exclude $M_{max} - M_{min} + 1 = 0$?
10. Prove (C.26).
11. Prove (C.29).
12. Derive (C.35).
13. Show that I_+J_- commutes with $(I_z + J_z)$. Use (C.10).

Appendix D

The ground levels and their electron configurations for neutral atoms. (The number of electrons in each shell is specified in the table.)

Z \ nl	1s	2s	2p	3s	3p	3d	4s	4p	4d	4f	...	Ground level
1 H	1											$^2S_{1/2}$
2 He	2											$^1S_0$
3 Li		1										$^2S_{1/2}$
4 Be		2										$^1S_0$
5 B		2	1									$^2P_{1/2}$
6 C	full	2	2									$^3P_0$
7 N		2	3									$^4S_{3/2}$
8 O		2	4									$^3P_2$
9 F		2	5									$^2P_{3/2}$
10 Ne		2	6									$^1S_0$
11 Na				1								$^2S_{1/2}$
12 Mg				2								$^1S_0$
13 Al				2	1							$^2P_{1/2}$
14 Si	full	full		2	2							$^3P_0$
15 P				2	3							$^4S_{3/2}$
16 S				2	4							$^3P_2$
17 Cl				2	5							$^2P_{3/2}$
18 Ar				2	6							$^1S_0$
19 K				2	6		1					$^2S_{1/2}$
20 Ca				2	6		2					$^1S_0$
21 Sc				2	6	1	2					$^2D_{3/2}$
22 Ti				2	6	2	2					$^3F_2$
23 V				2	6	3	2					$^4F_{3/2}$
24 Cr				2	6	5	1					$^7S_3$
25 Mn				2	6	5	2					$^6S_{5/2}$
26 Fe				2	6	6	2					$^5D_4$
27 Co	full	ful!		2	6	7	2					$^4F_{9/2}$
28 Ni				2	6	8	2					$^3F_4$
29 Cu				2	6	10	1					$^2S_{1/2}$
30 Zn				2	6	10	2					$^1S_0$
31 Ga				2	6	10	2	1				$^2P_{1/2}$
32 Ge				2	6	10	2	2				$^3P_0$
33 As				2	6	10	2	3				$^4S_{3/2}$
34 Se				2	6	10	2	4				$^3P_2$
35 Br				2	6	10	2	5				$^2P_{3/2}$
36 Kr				2	6	10	2	6				$^1S_0$

Z \ nl	1s	2s, p	3s, p, d	4s 4p 4d 4f	5s 5p 5d 5f 5g	Ground level
37 Rb				2 6	1	$^2S_{1/2}$
38 Sr				2 6	2	$^1S_0$
39 Y				2 6 1	2	$^2D_{3/2}$
40 Zr				2 6 2	2	$^3F_2$
41 Nb				2 6 4	1	$^6D_{1/2}$
42 Mo				2 6 5	1	$^7S_3$
43 Tc				2 6 5	2	$^6S_{5/2}$
44 Ru				2 6 7	1	$^5F_5$
45 Rh	full	full	full	2 6 8	1	$^4F_{9/2}$
46 Pd				2 6 10		$^1S_0$
47 Ag				2 6 10	1	$^2S_{1/2}$
48 Cd				2 6 10	2	$^1S_0$
49 In				2 6 10	2 1	$^2P_{1/2}$
50 Sn				2 6 10	2 2	$^3P_0$
51 Sb				2 6 10	2 3	$^4S_{3/2}$
52 Te				2 6 10	2 4	$^3P_2$
53 I				2 6 10	2 5	$^2P_{3/2}$
54 Xe				2 6 10	2 6	$^1S_0$

Appendix D

Z	1s	2s,p	3s,p,d	4s	4p	4d	4f	5s	5p	5d	5f	5g	6s	Ground level
55 Cs				2	6	10		2	6				1	$^2S_{1/2}$
56 Ba				2	6	10		2	6				2	$^1S_0$
57 La				2	6	10		2	6	1			2	$^2D_{3/2}$
58 Ce				2	6	10	mixed configurations							$J = 4$
59 Pr				2	6	10	3	2	6				2	$^4I_{9/2}$
60 Nd				2	6	10	4	2	6				2	$^5I_4$
61 Pm				2	6	10	5	2	6				2	$^6H_{5/2}$
62 Sm				2	6	10	6	2	6				2	$^7F_0$
63 Eu				2	6	10	7	2	6				2	$^8S_{7/2}$
64 Gd				2	6	10	7	2	6	1			2	$^9D_2$
65 Tb				2	6	10	8	2	6	1			2	$^8G_{13/2}$
66 Dy				2	6	10	10	2	6				2	$^5I_8$
67 Ho				2	6	10	11	2	6				2	$^4I_{15/2}$
68 Er				2	6	10	12	2	6				2	$^3H_6$
69 Tm				2	6	10	13	2	6				2	$^2F_{7/2}$
70 Yb	full	full	full	2	6	10	14	2	6				2	$^1S_0$
71 Lu								2	6	1			2	$^2D_{3/2}$
72 Hf								2	6	2			2	$^3F_2$
73 Ta								2	6	3			2	$^4F_{3/2}$
74 W								2	6	4			2	$^5D_0$
75 Re								2	6	5			2	$^6S_{5/2}$
76 Os								2	6	6			2	$^5D_4$
77 Ir								2	6	7			2	$^4F_{9/2}$
78 Pt					full			2	6	9			1	$^3D_3$
79 Au								2	6	10			1	$^2S_{1/2}$
80 Hg								2	6	10			2	$^1S_0$

nl Z	$1s$	$2s,p$	$3s,p,d$	$4s,p,d,f$	$5s$	$5p$	$5d$	$5f$	$5g$	$6s$	$6p$	$6d$	$7s$	Ground level
81 Tl					2	6	10			2	1			$^2P_{1/2}$
82 Pb	full	full	full	full	2	6	10			2	2			$^3P_0$
83 Bi					2	6	10			2	3			$^4S_{3/2}$
84 Po					2	6	10			2	4			$^3P_2$
85 At					2	6	10			2	5			$^2P_{3/2}$
86 Rn					2	6	10			2	6			$^1S_0$
87 Fr					2	6	10			2	6		1	$^2S_{1/2}$
88 Ra					2	6	10			2	6		2	$^1S_0$
89 Ac					2	6	10			2	6	1	2	$^2D_{3/2}$
90 Th					2	6	10			2	6	2	2	$^3F_2$
91 Pa					2	6	10	2		2	6	1	2	$^4K_{11/2}$
92 U					2	6	10	3		2	6	1	2	$^5L_6$
93 Np					2	6	10	4		2	6	1	2	$^6L_{11/2}$
94 Pu					2	6	10	6		2	6		2	$^7F_0$
95 Am	full	full	full	full	2	6	10	7		2	6		2	$^8S_{7/2}$
96 Cm					2	6	10	7		2	6	1	2	$^9D_2$
97 Bk					2	6	10	8?		2	6	1?	2	$^8G_{15/2}$?
98 Cf					2	6	10	10?		2	6		2	$^5I_8$?
99 E					2	6	10	11		2	6		2	$^4I_{15/2}$
100 Fm					2	6	10	12		2	6		2	$^3H_6$
101 Mv					2	6	10	13		2	6		2	$^2F_{7/2}$
102 No					2	6	10	14		2	6		2	$^1S_0$
103 Lw					2	6	10	14		2	6	1	2	$^2D_{3/2}$

Appendix E

Units

This book is written in SI units and, where appropriate, the 'pet' units of the specialist have been used, e.g., Å and cm^{-1} (optical spectroscopy), mK (high-resolution optical spectroscopy), MHz or Hz (radiofrequency spectroscopy). One should be able to translate these units into c.g.s. units, for it is one of the purposes of the book to introduce the reader to a vast body of literature which is written in a language of traditional units and special notations.

But the matter does not end here, for there is another set of units, to be found particularly in theoretical work, which is much more appropriate to the subject of atomic structure. The units are called atomic units (a.u.). and they are usually based on the non-rationalized c.g.s. system in which, on the one hand the quantities $4\pi\varepsilon_0$ and $\mu_0/4\pi$ do not appear, but on the other hand the speed of light is involved in the conversion from electromagnetic to electrostatic c.g.s. units. In the system of atomic units certain dimensional constants are set equal to unity so that they conveniently disappear from many formulae. These constants are

Rest mass of the electron

$$m_0 \equiv 1 \text{ a.u. of mass,} \qquad (E.1)$$

Magnitude of the charge of the electron

$$e \equiv 1 \text{ a.u. of charge,} \qquad (E.2)$$

Radius of the first Bohr orbit

$$a_0 = \frac{\hbar^2}{m_0 e^2} \equiv 1 \text{ a.u. of length.} \qquad (E.3)$$

(Note that the formulae in this appendix are written in c.g.s. form.) Derived from these we have

$$\hbar = (m_0 e^2 a_0)^{1/2} = 1 \text{ a.u. of action (or of angular momentum).} \quad (E.4)$$

The dimensionless fine-structure constant, which is of central importance in electromagnetic interactions, is

$$\alpha = \frac{e^2}{\hbar c} \approx \frac{1}{137},$$ (E.5)

so that the speed of light in atomic units is

$$c = \alpha^{-1} \approx 137 \text{ a.u. of speed.}$$ (E.6)

Thus 1 a.u. of speed is αc, which is the speed of the electron in the first Bohr orbit of hydrogen according to the Bohr model. The unit of time is the time taken to travel a distance a_0 with speed αc:

$$\tau_0 \equiv \frac{a_0}{\alpha c} = \frac{\hbar^3}{m_0 e^4} = 1 \text{ a.u. of time.}$$ (E.7)

The unit of energy is chosen to be

$$e^2/a_0 = 1 \text{ a.u. of energy,}$$ (E.8)

so the ionization energy of hydrogen, hcR_∞, is $\frac{1}{2}$ a.u. (Notice that this choice of energy unit is not quite universal, for some authors have used $\frac{1}{2}e^2/a_0$ as the unit.)

One of the consequences of using atomic units is that Schrödinger's equation describing the gross structure of hydrogen reads

$$\left(-\tfrac{1}{2}\nabla^2 - \frac{Z}{r} \right)\psi = E\psi$$ (E.9)

in which the constants $\hbar$, m_0, and e^2 no longer appear. Notice that atomic units are not the same as relativistic units, in which the rest energy of the electron, $m_0 c^2$, is set equal to 1. In atomic units this energy is of magnitude α^{-2}.

We list below, in atomic, c.g.s. and SI units, accurate values of some fundamental constants (table E.1) and approximate values of some further quantities derived from the fundamental constants (table E.2).

Table E.1

Accurate values of some physical constants in atomic, c.g.s. and SI units. The numbers in parentheses are the standard-deviation uncertainties in the last digits of each quoted value. These values are selected from table 33.1 of the paper by E. R. Cohen and B. N. Taylor, *J. Phys. Chem. Ref. Data* **2**, 663, 1973. In column 1 the quantities in square brackets multiply the c.g.s. expressions to convert them into SI expressions ($\varepsilon_0 = 10^7/(4\pi c^2) \approx 8.85 \times 10^{-12}$ F m^{-1}).

Quantity	Symbol	Value in a.u.	Value	in: c.g.s. units	SI units
Electron rest mass	m_0	1	9.109534(47)	10^{-28} g	10^{-31} kg
Magnitude of electron charge	e	1	1.6021892(46)	10^{-20} e.m.u.	10^{-19} C
			4.803242(14)	10^{-10} e.s.u.	
Planck's constant	h	2π	6.626176(36)	10^{-27} erg s	10^{-34} J s
	$\hbar$	1	1.0545887(57)	10^{-27} erg s	10^{-34} J s
Fine-structure constant $\dfrac{e^2}{\hbar c}\left[\dfrac{1}{4\pi\varepsilon_0}\right]$	α	7.2973506(60)	7.2973506(60)	10^{-3}	10^{-3}
	α^{-1}	137.03604(11)	137.03604(11)	10^0	10^0
Speed of light	c	α^{-1}	2.997924580(12)	10^{10} cm s^{-1}	10^8 m s^{-1}
Electron charge to mass ratio	e/m_0	1	1.7588047(49)	10^7 e.m.u. g^{-1}	10^{11} C kg^{-1}
			5.272764(15)	10^{17} e.s.u. g^{-1}	
Ratio of proton mass to electron mass	M/m_0	1836.15152(70)	1836.15152(70)	10^0	10^0
Radius of first Bohr orbit $\dfrac{\hbar^2}{m_0 e^2}[4\pi\varepsilon_0]$	a_0	1	5.2917706(44)	10^{-9} cm	10^{-11} m
Rydberg constant $\dfrac{me^4}{4\pi\hbar^3 c}\left[\dfrac{1}{4\pi\varepsilon_0}\right]^2$	R_∞	$\alpha/2$	1.097373177(83)	10^5 cm^{-1}	10^7 m^{-1}
Bohr magneton $\dfrac{e\hbar}{2m_0 c}[c]$	μ_B	$\alpha/2$	9.274078(36)	10^{-21} erg G^{-1}	10^{-24} J T^{-1}
Electron magnetic moment in Bohr magnetons	μ_e/μ_B	1.0011596567(35)	1.0011596567(35)	10^0	10^0

222

Table E.2

Approximate values of some supplementary physical quantities in atomic, c.g.s., and SI units.

Quantity	Value in a.u.	Value	in: c.g.s. units	SI units
Atomic unit of energy $$\frac{e^2}{a_0}\left[\frac{1}{4\pi\varepsilon_0}\right] = \frac{me^4}{\hbar^2}\left[\frac{1}{4\pi\varepsilon_0}\right]^2 = 2hcR_\infty$$	1	4·36 27·2	10^{-11} erg eV	10^{-18} J eV
Atomic unit of time $$\tau_0 = \frac{a_0}{\alpha c} = \frac{\hbar^3}{m_0 e^4}\left[4\pi\varepsilon_0\right]^2$$	1	2·42	10^{-17} s	10^{-17} s
Rest energy of electron $$m_0 c^2$$	α^{-2}	8·19 5·11	10^{-7} erg 10^5 eV	10^{-14} J 10^5 eV
Atomic unit of electric dipole moment $$ea_0 = \frac{\hbar^2}{m_0 e}\left[4\pi\varepsilon_0\right]$$	1	2·54 8·48	10^{-18} e.s.u. cm	10^{-30} C m
Bohr magneton in frequency units $$\mu_B/h$$	$\alpha/4\pi$	1·400	Mc/s G^{-1}	10^4 MHz T^{-1}
Interaction energy (fine structure) $$\alpha^2\frac{e^2}{a_0}\left[\frac{1}{4\pi\varepsilon_0}\right]$$	α^2	2·32	10^{-15} erg	10^{-22} J
$$\alpha^2\frac{e^2}{\hbar}\frac{1}{a_0}\left[\frac{1}{4\pi\varepsilon_0}\right]$$	$\alpha^2/2\pi$	3·5	10^5 Mc/s	10^5 MHz
$$\alpha^2\frac{e^2}{hc}\frac{1}{a_0}\left[\frac{1}{4\pi\varepsilon_0}\right]$$	$\alpha^3/2\pi$	1·17	10^1 cm^{-1}	10^3 m^{-1}

Magnetic field corresponding to an energy $\alpha^2 e^2/a_0$

$$B = \frac{\alpha^2}{\mu_B}\frac{e^2}{a_0}\left[\frac{1}{4\pi\varepsilon_0}\right] = 2\alpha\,\frac{e}{a_0^2}\left[\frac{1}{4\pi\varepsilon_0 c}\right] = \frac{4\mu_B}{a_0^3}\left[\frac{1}{4\pi\varepsilon_0 c^2}\right]$$

2α	2·5	10^5 G	10^1 T

Interaction energy (hyperfine structure)

$$\frac{m_0}{M}\alpha^2\frac{e^2}{a_0}\left[\frac{1}{4\pi\varepsilon_0}\right] = 4\frac{\mu_B\mu_N}{a_0^3}\left[\frac{1}{4\pi\varepsilon_0 c^2}\right]$$

$$= 4\frac{\mu_B\mu_N}{ha_0^3}\left[\frac{1}{4\pi\varepsilon_0 c^2}\right]$$

$$= 4\frac{\mu_B\mu_N}{hca_0^3}\left[\frac{1}{4\pi\varepsilon_0 c^2}\right]$$

$\dfrac{m_0}{M}\alpha^2$	1·26	10^{-18} erg	10^{-25} J
$\dfrac{m_0}{M}\dfrac{\alpha^2}{2\pi}$	1·9	10^2 Mc/s	10^2 MHz
$\dfrac{m_0}{M}\dfrac{\alpha^3}{2\pi}$	6·4	10^{-3} cm^{-1}	10^{-1} m^{-1}
	6·4	mK	

Magnetic field corresponding to an energy $\dfrac{m_0}{M}\alpha^2\dfrac{e^2}{a_0}$

$$B = 4\frac{\mu_N}{a_0^3}\left[\frac{1}{4\pi\varepsilon_0 c^2}\right]$$

$2\dfrac{m_0}{M}\alpha$	1·36	10^2 G	10^{-2} T

Index